OrCAD® PSpice®
with Circuit Analysis

Third Edition

Franz J. Monssen

Prentice
Hall

Upper Saddle River, New Jersey
Columbus, Ohio

Library of Congress Cataloging-in-Publication Data

Monssen, Franz.

    OrCAD PSpice with circuit analysis/ Franz Monssen.—3rd ed.

      p. cm

    Rev. ed. of: MicroSim PSpice with circuit analysis. 2nd ed. c1998.

    Includes index.

    ISBN 0-13-017035-6 (alk. paper)

    1.   Electronic circuit analysis—Data processing. 2. PSpice. I Monssen, Franz.

    MicroSim PSpice with circuit analysis. II. Title.

    TK 454.M65 2001

    621.319'2'07855369—dc21   00-061975

Vice President and Publisher: Dave Garza
Editor in Chief: Stephen Helba
Acquisitions Editor: Scott J. Sambucci
Production Editor: Rex Davidson
Design Coordinator: Karrie M. Converse-Jones
Cover Designer: Thomas Borah
Cover Photo: SuperStock
Production Manager: Pat Tonneman
Marketing Manager: Ben Leonard

The book was printed and bound by Victor Graphics, Inc.
The cover was printed by Victor Graphics, Inc.

OrCAD® and PSpice® are registered trademarks of Cadence Design Systems.

Second edition, entitled *MicroSim® PSpice® with Circuit Analysis*, copyright 1998 Prentice-Hall, Inc.

First edition, entitled *PSpice® with Circuit Analysis*, copyright 1993 by Macmillan Publishing Company.

10 9 8 7 6 5 4 3 2 1

ISBN 0-13-017035-6

PREFACE

Since the publication of the first edition of *PSpice with Circuit Analysis*, the change-over from the DOS to the Windows format and the proliferation of relatively cheap personal computers have created the need for this third edition of the text.

The impact of such programs as PSpice will be profound in the workplace of the present and future electrical engineer. Also, they will increasingly affect the fashion in which electrical engineering and electronics technology are taught in colleges and technical schools. The analysis of electrical and electronic circuits even of modest size involves both complex and lengthy calculations. By means of PSpice, circuit complexity is far less a hindrance to a successful analysis of electrical circuit behavior. Relatively few rules of program syntax together with a few click-and-drag operations allow the electrical engineer and the student to solve complex circuits and produce circuit schematics of professional quality.

The successful evaluation of a formula, when done by a calculator, gives the relationship between circuit variables at a particular operating point. PSpice, by contrast, allows for a global perspective of circuit behavior. The ease, compared to traditional methods, by which a frequency or transient (time) analysis can be performed well illustrates the point. Oscilloscope-like displays of circuit variables, their mathematical relationships, and concepts such as the RMS value of a signal can all be accurately and quickly displayed.

PSpice allows a shift of emphasis away from computation of circuit variables toward their interpretations. It also allows a shift away from the analysis on the component level of circuits to the analysis of systems consisting of many circuits. Traditionally, students spend considerable time analyzing circuits containing a single bipolar transistor. However, practical circuits such as multistage amplifiers, operational amplifiers, active filters, and communication circuits all contain many transistors in addition to numerous other solid-state devices. Circuits of such complexity can be analyzed with relative ease by the PSpice program.

Educators are challenged to incorporate these PSpice capabilities into the existing curriculum. With the pedagogical approach in this text, a particular circuit phenomenon is observed by means of graphical and numerical output data generated by PSpice. Having observed the phenomenon, the student is prompted to seek an explanation. At this point, mathematical concepts and formulas are introduced. The student is asked to show the correlation between data and calculation. A typical example of this approach is found in Chapter 6 of this text. There, the response of a RC circuit to a sinusoidal voltage is investigated by means of the PSpice program. Subsequent to that, the phasor method is introduced to solve for the circuit response. Finally, the results of that method and the PSpice data are compared. The structure of this book reflects this approach. Although its topical outline supports the traditional electrical engineering curriculum, a dual approach is taken. Every new circuit concept is introduced with its relevant PSpice commands. Subject matter and the PSpice program are used in a mutually supportive fashion.

This text is neither an electrical engineering textbook with the PSpice program relegated to an appendix nor is it simply a reference manual for PSpice. The underlying assumption of this book is that the combination of subject matter and the PSpice program can foster conceptual understanding and competence to advance the learning process. This text does not advocate the neglect of tradition teaching methods in favor of the use of the PSpice program; rather, it uses them in conjunction with the program.

The material in this book is divided into ten chapters, which cover the topics usually found in a one-year course in electrical circuit analysis. Chapter 1 is devoted to the analysis of dc circuits containing single and multiple independent current and voltage sources. The analysis of circuits containing dependent current and voltage sources is also included in this chapter. These devices play an important role in the modeling of many solid-state devices.

Chapter 2 introduces some fundamental network theorems as applied to dc circuits. The Superposition theorem, Thevenin's theorem, and Norton's theorems are demonstrated. Source conversions and the Maximum Power Transfer theorems conclude the chapter.

Chapter 3 introduces the reader to the transient phenomena encountered in RC and RL circuits. The continuity of a capacitor's voltage and an inductor's current are related to the power flow and the energy content of these elements. The concept of the time constant is introduced. The circuit response to a linear pulse train is introduced. The model of a switch as used in PSpice is applied to circuits.

Chapter 4 has as its subject the application of sinusoidal currents and voltage to resistive circuits. The latter are used so as not to introduce unnecessary complications at this time. The effects of time and phase shifts, their relation, and their effect on circuit behavior are studied. The sums of sinusoidal currents and voltages are obtained.

Chapter 5 investigates the steady-state sinusoidal response of series, parallel RC, RL, and RLC circuits. The concept of impedance is demonstrated graphically by using PROBE plots. These plots demonstrate that admittance and impedance are steady-state concepts. The equivalency between various series and parallel circuits is demonstrated.

Chapter 6 investigates the total response of series and parallel RC, RL, and RLC circuits. This response is shown to be the sum of the forced response due to the source and the transient, or natural, response of a circuit. The subject of electrical resonance investigated.

Chapter 7 extends the concepts of Chapter 2 to circuits containing resistors, capacitors, and inductors. Superposition is applied to circuits containing both ac and dc current and voltage sources. The ability of students to analyze such circuits will become increasingly important as they progress to electronic circuits containing solid state devices. It is demonstrated that Norton and Thevenin impedances are generally complex quantities. It is shown that maximum power from a source to a load will flow when the source impedance is the complex conjugate of the load impedance.

Chapter 8 shows that power and energy flow in alternating current circuits. The power and energy relations for resistors, capacitors, and inductors are investigated. The concept of apparent, real, and reactive power is introduced and obtained by PSpice. It is important to note that this chapter provides the theoretical basis upon which the electrical utility industry is built.

Chapter 9 introduces the reader to the frequency response of RC, RL, and RLC circuits. The necessary PSpice statements are introduced and applied. Both magnitude and phase plots are obtained and the impedance of circuits is plotted as a function of frequency. The concepts of the 3dB frequency are introduced and applied to the filters. The RC circuit is used both in its high-pass and low-pass configurations. The effect of multiple sections on roll-off is demonstrated. The characteristics of a band-pass filter are examined. The concept of the Q factor is covered. The relationship between transient analysis and AC (frequency) analysis is stressed.

Chapter 10 applies non-sinusoidal current and voltage sources to electrical circuits. The ability of the PSpice program to perform a Fourier analysis is demonstrated. The Fourier transforms of some standard waves are obtained. The PSpice program calculates the total harmonic distortion (THD) of a wave. The RMS value of a wave train and the power it delivers to a load are calculated. The effects of half-wave and full-wave rectification on a Fourier spectrum are investigated. Wave symmetry and the effects of time shift of a wave are explored. Finally, a square pulse is applied to a RC high-pass filter and a band-pass filter. The effects of these circuits on the Fourier transform are investigated.

Acknowledgments

At the completion of this book, it is proper to thank those who contributed to it. Even a short reflective pause makes the author aware of his indebtedness to so many. It is impossible to tell how many years ago that the seed of this book was planted. There were the dedicated teachers at the City College of New York who introduced a young and untutored mind to the exciting world of electrical engineering. There are those by whose prior intellectual efforts this author has profited. By formal lecture and informal discussions over a cup of coffee, remarks were made and insights gained that found expression in this book.

There are those with whom the author has had the privilege and pleasure of a congenial professional relationship. Among them is the late Professor Joseph Aidala, the former chairperson of the Electrical and Computer Engineering Technology Department at Queensborough Community College. Thanks are due to the present chairperson, Dr. Louis Nashelsky, whose professional competence is surpassed only by his human decency. Thanks is due to Professor Robert Boylestad, friend and colleague, for introducing me to various people at Prentice Hall who became interested in the publication of a book on PSpice.

The work not only has to be written but it also needs to be published. This obliges me to thank all those involved in that part of the process. Initial thanks must go to Scott Sambucci, the Electronics Technology acquisitions editor at Prentice Hall. A special thanks goes to Rex Davidson, who guided the editing process through its various stages. Finally, a thank you to Linda Thompson, who did the editing and added to my humility.

Writing on a more personal note, a special thanks must go to my dearest Daphine, who wanders with me along life's road. Thanks also to my two sons, Georg (without the final e) and Stefan, who, despite having been teenagers, turned into two wonderful adults.

CONTENTS

3

TRANSIENTS IN RC AND RL CIRCUITS 71

4

SINUSOIDAL WAVEFORMS IN RESISTIVE CIRCUITS 107

5

STEADY STATE SINUSOIDAL RESPONSES OF RC, RL AND RLC CIRCUITS 139

6

TOTAL RESPONSE OF RC, RL AND RLC CIRCUITS WITH SINUSOIDAL SOURCES   179

7

ALTERNATING CURRENT NETWORK THEOREMS   203

8

POWER AND ENERGY IN ALTERNATING CURRENT CIRCUITS   235

9

FREQUENCY RESPONSE OF RC, RL AND RLC CIRCUITS 271

10

CIRCUITS WITH NONSINUSOIDAL SOURCES 327

PSPICE ANALYSIS OF DC CIRCUITS

ANALYSIS OF A SERIES CIRCUIT

Kirchhoff's Voltage Law for Series Circuits

A series circuit has only one current path. If a voltage or current source is connected to a series circuit, the current is the same throughout the circuit. For such a circuit, Kirchhoff's voltage law states: "the algebraic sum of the voltages around the closed path of the circuit is equal to zero volts." It follows that the electrical power delivered by the source(s) is equal to the power consumed by the resistor(s). This is a demonstration of the law of the conservation of energy as applied to electrical circuits.

Ideal DC Current and Voltage Sources

An ideal dc voltage source maintains a constant voltage across its terminals independent of time and any external circuit elements connected to it. An ideal dc current source maintains a constant current between its terminals independent of time and any external circuit elements connected to it.

The Resistor

We begin our analysis with purely resistive circuits. A resistor dissipates electrical energy over time. The rate of energy dissipation is defined as the power to the resistor. The ratio of the voltage across and the current through a resistor is defined as the resistance of a resistor. Its units are most commonly ohms (Ω) or kilohms ($k\Omega$).

Preparing the Circuit Schematic

New file

In the analysis that follows, **OrCad's PSpice Evaluation Software Release 9.1** is used. We start our analysis with the creation of the circuit schematic. Our circuit consists of three resistors in series. Their resistances are 1 $k\Omega$, 4 $k\Omega$ and 5 $k\Omega$. A 20 volt dc voltage source is connected to the circuit. Do the following:

1. On the computer's **Desktop** screen, click on **Start,** move cursor right to **Program,** move right to **OrCad Demo**, move right to **Capture CIS Demo** and click on it. The **OrCad Capture** screen opens.
2. Click on **File**, move to **New**, move to **Project** and click; the **New Project** dialog box opens.
3. In the **Name** box, enter the name of circuit. The author used Figure 101.

4.   In the **Location** box, type information about where circuit file will reside.
5.   Select **Analog or Mixed-Signal Circuit Wizard** by clicking on it.
6.   Click on **OK**. The **Analog Mixed-Mode Project Wizard** dialog box opens.
7.   Select analog.olb. Click on **Finish**. The **OrCad Capture [I-(Schematic1: Page 1)]**
     appears. If toolbar does not appear on right edge of the Schematic page, click
     anywhere on the screen. The screen is shown next.

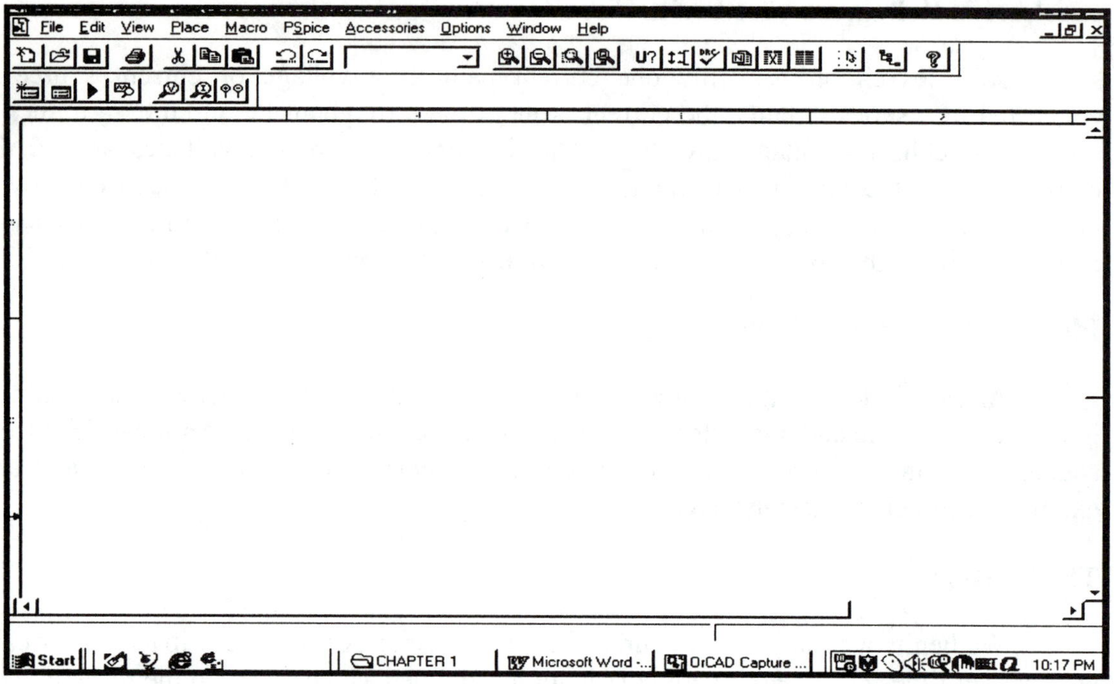

Existing file
     If an existing file is to be recalled for simulation:
1.   From **Desktop**, click on **Start**, drag to **OrCad Demo**, drag to **Capture CIS Demo**, and
     click on it. The **OrCad Capture** screen opens.
2.   Click on **File**, drag to **Open**, drag to **Project** and release. The **Open Project** dialog box
     opens.

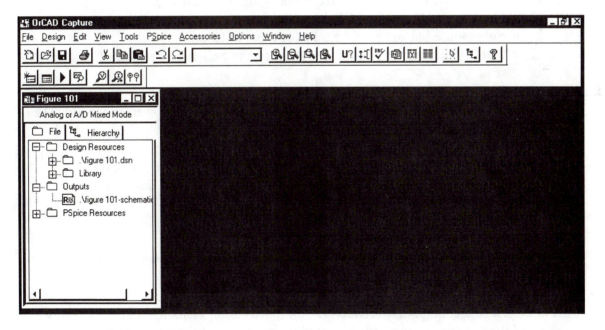

3. Click on the desired file and click on **Open**. The **OrCad Capture** screen opens.
4. Click on the plus sign [+] on the left side of selected file.
5. Click on the plus sign [+] on the left side of **Schematics1** folder. The **Page1** icon opens.
6. Double click on the **Page1** icon. The **OrCad Capture-[(Schematics]:Page1)]** opens. It contains the desired circuit schematic.

Placement of Parts

Getting a resistor
1. Click on the icon with diode legend. It is the uppermost icon in the task bar on the right edge of the Schematic Page. The legend **Place Part** will appear briefly.
2. Click on it. The **Place Part** dialog box will appear.
3. In **Libraries** box, click on **ANALOG**.
4. In **Part** box, type **R** for resistor or scroll to **R;** click on it to select it.
5. In **Place Part** box, click on **OK**. The symbol of a resistor will appear on the Schematic Page.
6. Drag it to the desired location. Click to place it.
7. Drag to the next location to place the second resistor. Click to place it.
8. Repeat procedure for the third resistor.
9. Click the right mouse button. Select **End Mode** and click on it.
10. Deselect the third resistor by placing the pointer anywhere on the **Schematic** page and click.

Getting a voltage source
1. Click on the **Place Part** icon. The **Place Part** dialog box will appear.
2. In the **Libraries** box, scroll to and select **SOURCE**.
3. In the **Part** box, scroll to and select **VDC** by clicking on it. Click on **OK**.
4. Drag **VDC** to the desired location. Click to place it.
5. Click the right mouse button. Select **End Mode** and click on it.
6. Deselect **VDC** by placing the pointer anywhere on the **Schematic** page and click on it.

Wiring the Components
1. On the right side of screen, on the tool bar, point to the second icon from the top. The legend **Place Wire** will appear briefly. Click on it.
2. Drag the cross-hair pointer to positive end of **VDC** source. Click on it
3. Move the cross-hair pointer to one terminal marker of first resistor. Click on it.
4. Move the cross-hair pointer to other terminal marker of resistor. Click on it.
5. Move onto the terminal marker of next resistor. Click on it.
6. Repeat the procedure until all components are connected.
7. Click the right mouse button. Select **End Wire** and click on it.
8. Deselect the last wire segment by clicking anywhere on the **Schematic** page.

Labeling the nodes
   A node is a connecting point between two or more circuit elements In our circuit, we can think of the connecting wires as elongated nodes. To identify the nodes, we shall label each of them.
1. Click on a wire to select it.
2. Point to the third icon from the top. It has the legend **N1**. Click on it.
3. The **Place Net Alias** dialog box appears.
4. In that box, type in the desired label. Click on **OK**.

5. Drag the small rectangle to selected node (wire). Click on it. The selected label will appear.
6. Click the right mouse button to select **End Mode**. Click to end labeling of the node.
7. Repeat for all other nodes.
8. Important: one node must be labeled zero! It is the ground node.

Setting the resistance values

All resistors placed on the **Schematic** page initially have a value of 1 kΩ. To change these values, if needed, proceed as follows:
1. Double click on the resistance value of a selected resistor. The **Display Properties** box appears.
2. In the **Value** box, type in the desired resistance. Click on **OK**. Click to deselect.
3. Repeat for all resistors.

Renaming resistors

Should it be desired to rename the resistors, proceed as follows:
1. Double click on the label of resistor, for example, **R1**. The **Display Properties** box will appear.
2. In the **Value** box, type the desired name. Click on **OK**. The desired name will appear in place of the former name of the resistor.
3. Click anywhere on the screen to end process.

Setting the Voltage of VDC

We next set the voltage of **V1** to the desired 20 volts.
1. Double click on the voltage of V1. The **Display Properties** dialog box appears.
2. In the **Value** box, type 20V. Click on **OK**.
3. In **Schematic** page, click to deselect **V1**. Its new voltage of 20 volts will appear.

Placing Text on the Schematic page
1. Click on the **A** icon. It is at the bottom of the task bar. The **Place Text** box opens.
2. In the box, type the desired text. Click on **OK**.
3. Drag the rectangle to the desired location on **Schematic** page. Click to place.
4. Click on the right mouse button and select **End Mode**. Click on it.
5. Click in **Schematic** page to deselect text.

At the completion of this section the circuit schematic will appear as shown in Figure 1.01. We shall next prepare our circuit so that its voltages and its current can be obtained and displayed both in numerical and in graphic format.

Figure 1.01

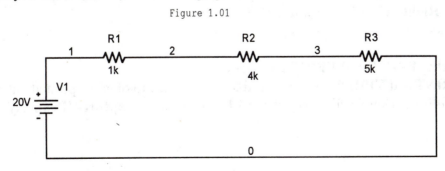

Obtaining the Numerical Output Data

During simulation, **PSpice** generates an **Output File** which checks the correctness of the simulations and generates an error message should there be an error in the syntax of the circuit file. It prints out the value of specified circuit variables.

The circuit voltage: VPRINT1

To print the voltage of any node with respect to ground, we place **VPRINT1** at the node of interest. It is desired to obtain the printed values of the voltages at nodes 2 and 3.

1. Click on the **Place part** icon.
2. In the **Place Part, Libraries** box, select **SPECIAL**.
3. Scroll to **VRPINT1,** click on it and click on **OK.**
4. Drag **VPRINT1** to node 2 and click to place it.
5. Drag **VPRINT1** to node 3 and click to place it.
6. Click Right, select **End Mode** and click on it.
7. Deselect **VPRINT1** by clicking anywhere on the **Schematic** page.

The circuit voltage: VPRINT2

To print the voltage between any two nodes in the circuit, place **VPRINT2** at the nodes of interest. It is desired to obtain the printed value of the voltage across **R2**.

1. Click on the **Place Part** icon.
2. In **Place Part, Libraries** box, select **SPECIAL**
3. Scroll to **VPRINT2**, click on it and click on **OK**. Note: **VPRINT2** has two leads that must be wired into the circuit.
4. Drag **VPRINT2** to a place above **R2**. Click to place it and wire it into the circuit.
5. Click on **Right**, select **End Mode** and click on it.
6. Deselect **VPRINT2** by clicking anywhere on the **Schematic** page.

The circuit current: IPRINT

Since the current in this circuit is everywhere the same, the **IPRINT** device can be placed at a convenient location. It is entered into a circuit like an ammeter.
1. Select the horizontal section of node 0 and press **Del** button.
2. Click on the **Place part** icon.
3. In the **Place Part, Libraries** box, select **SPECIAL.**
4. Scroll to **IPRINT,** click on it and click on **OK.**
5. Drag **IPRINT** to the deleted portion of node 0. Click to place it and rewire.
6. Click on **Right**, select **End Mode** and click on it.
7. Rewire node 0.

Activating the IPRINT and VPRINT devices

The **IPRINT** and **VPRINT** devices need to be activated to obtain a printed output. Proceed as follows: Double click on **VPRINT1**. The **OrCad Capture- [Property Editor]** page opens.

1. Scroll to the right until the **DC** box appears. Click on it and type **Y** in it.
2. Click on **Apply.**
3. Repeat the procedure for the other **VPRINT1**, **VPRINT2** and **IPRINT** devices.

At the end of these procedures, the schematic of Figure 1.01 will appear as shown here.

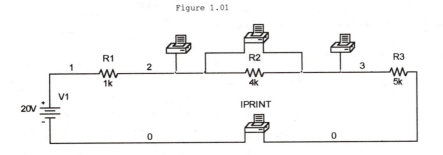

Figure 1.01

Running the Simulation

Setting theAnalysis Type and Sweep Parameters
1. Click on **PSpice**, and the **New Simulation** dialog box opens.
2. In its **Name** box, type 1. There may be several simulations for this circuit.

3. Click on **Create**. The **Simulations Settings** dialog box opens.
4. In **Analysis type** box, scroll to **DC Sweep**.
5 . In **Sweep variable** box, click on **Voltage source** to select it.
6. In **Name** box, type V1.
7. Retain **Linear** in **Sweep type**.
8. In **Start value** box, type 20 V
9. In **End value** box, type 20 V.
10. In **Increment** box, type 1 V.
11. Click on **OK** to accept all values.
     Note: it is not necessary to type the units in any of the last three boxes. The entry into the **Increment** box may appear to be confusing: it is best thought of as a multiplier constant. The **Simulation Settings** dialog box with all needed values entered is shown .

Start the Simulation
1. From the File menu, click on **Save**.
2. From the **PSpice** menu, click on **Run** to start the simulation.

Viewing the Results
The **PROBE** plot
Voltages and the circuit current
1. The **PROBE** plot. At the completion of the simulation, the **PROBE** screen shown opens.
2. Click on **Trace.** Clicking on **Add Trace** opens the **Add Traces** dialog box.
3. Click on a variable to select it. The author selected V(1). The expression for the selected variable will appear in the **Trace Expression** box.
4. Click on **OK** and the trace of V(1) will appear on the **PROBE** screen.
5. Repeat the process for other variables.

The author selected the voltage traces of V(1), V(2) and V(3). Their values are 20 volts, 18 volts and 10 volts. They are shown on the **PROBE** plot.

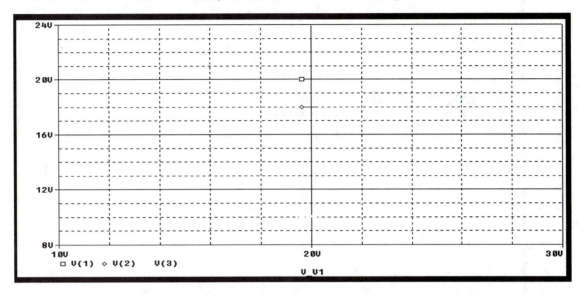

The X-axis of the **PROBE** displays the voltage of **V1** as the independent variable. The voltages V(1), V(2) and V(3) are hardly visible. Let us change the X-axis to display time. To do this, we shall perform a **transient analysis** of the circuit. In the syntax of the **PSpice** program, the term **transient** means a time analysis. This is not to be confused with a **transient analysis** in classical differential equation analysis. In that, the term means the perturbation in time of circuit variables as the circuit adjust from an initial to a new energy level.

We proceed as follows:
1. Open **Figure 101 Schematic, Page 1.**
2. Click on **Spice** and drag and click on **New Simulation Profile**.
3. In the **New Simulation** dialog box, enter a number for the simulation.
4. Click on **Create.** The **Simulation Setting** dialog box opens.
5. In Analysis type: select **Time Domain(Transient).**
6. Click in **Run to time**: type 10s. This will give a 10 second time sweep. Since this is a dc circuit, any arbitrary time value would serve as well.
7. Click on **Maximum step size** box and type 1s. For this dc circuit, the step size is not critical.
8. Click on **OK** to accept all values. The **Simulation Settings** dialog box will appear as shown.

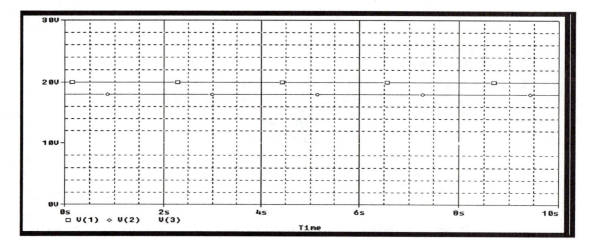

The results of the Time domain (**Transient**) analysis are shown on the **PROBE** plot. The traces for the voltages V(1), V(2) and V(3) were selected The appearances of the traces are those one would see on an oscilloscope. They show that all circuit voltages are independent of time. This is true of dc currents and voltages.

We observe the value of the circuit current by selecting the traces I(R1), I(R2) and I(R3).   Their values are all 2 milliamps as it must be for a series circuit. The current through the voltage source **V1** is important. Its trace is I(V1) and it has a value of − 2 milliamps. This is because of the syntax of the **PSpice** program. Positive current will enter the positive terminal of a device. It is receiving power. Negative current leaves the positive terminal of a device. The device is losing power. The power received by the resistors from source **V1** is equal to the power lost by **V1**. The current traces are shown next.

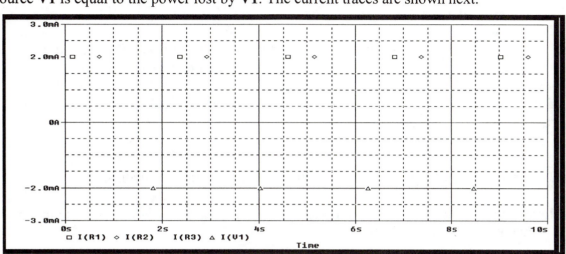

The voltage source **V1** has an explicit plus sign, whereas that for the resistors is implicit. We shall see in Figure 1.02 that resistors can be rotated. The syntax of the **PSpice** program dictates that all resistors rotate around their positive terminal.

Power in the Circuit

PROBE allows for the plotting of power.   Power plots are obtained by forming the product of the voltage across a device multiplied by the current through that device.   The expressions for the power of each circuit element are shown at the bottom of the screen. The voltage differences across the resistors are enclosed within parentheses.

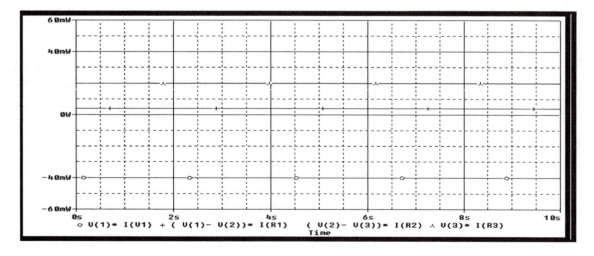

V1 is delivering 40 milliwatts of power to the circuit, hence its minus sign. Resistors **R1**, **R2** and **R3** are receiving 4 milliwatts, 16 milliwatts and 20 milliwatts respectively, hence the plus signs. The algebraic sum of all the powers in the circuit is zero milliwatts, as was predicted. This result is required by law of the conservation of energy as applied to electrical circuits.

The Output File

During simulation, the **PSpice** program generates an **Output File**. To call for it, with the **PROBE** screen open, click on **View**, drag to **Output File** and release. The **Output File** opens. It contains a listing of the analysis used, the **Alias File** and the **PRINT** statements called for. To save space, only the voltages and the circuit current as printed in the file are shown.

V_V1	V(3)	V(2,3)	V(2)	I(V_PRINT5)
2.000E+01	1.000E+1	8.000E+00	1.800E+1	-2.000E-03

All voltages and the current are expressed in scientific notation. These values are the same as previously shown on their **PROBE** traces.

Analysis of a Resistive Voltage-Divider Circuit

The previous section was of considerable length because the reader needed to be introduced to every step in the creation and the analysis of the circuit in Figure 1.01 The reader is advised to consult it as needed. From now on, only new components, analysis types and procedures will be stated. These will be listed under the heading **What's New ?** On a special note: Figure 1.01 was entered as Figure 101 in the **Name** box of the **New Project** dialog box. If the decimal point had been used, the file name would have been truncated to Figure 1.

What's new?
1. A variable voltage source
2. Rotation of a component
3. The **Probe** cursors **A1** and **A2**

Our circuit consists of the two resistors shown connected to a variable dc voltage source **V1** that can change its voltage from zero volts to 18 volts in steps of 1 volt. It is desired to plot the voltage across **R2** as a function of **V1**, defined as the input voltage. The circuit is labeled **Figure 1.02.**

Figure 1.02

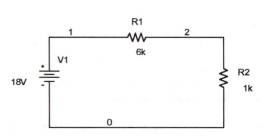

To rotate **R2** do the following:
1. Select **R2** as in the previous section and place it where desired.
2. Click to place; click the right mouse button to **End Mode**.

3. With **R2** still selected, depress the **Ctrl** button and simultaneously click on the **R** button. **R2** will be rotated 90 degrees counter-clockwise.
4. Repeat step 3 until **R2** is in a vertical position. Its positive terminal is the top one.
5. Complete the construction of the schematic as in the previous section.

Running the simulation
1. Perform all steps as in the previous section.
2. When the **Simulation Settings** dialog box opens, enter the values for **V1** as shown.
3. When completed, click on **OK**.
4. From **PSpice** menu, click on **Run** to start the simulation.

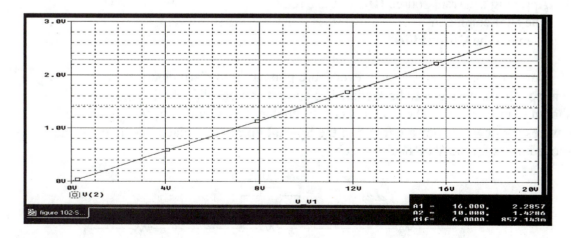

Viewing the analysis results

Activating the cursors

PROBE has two cursors **A1** and **A2**. They will operate on the first trace listed. To activate them do the following:
1. Click on the cursor icon, located two icons to the right of the icon labeled **ABC.**
2. Depress the right horizontal arrow key to move cursor **A1** to the right; depress the left horizontal arrow key to move cursor **A1** to the left.
3. Depress the Shift key and the right horizontal key to move cursor **A2** to the right; depress the Shift key and the left horizontal key to move cursor **A2** to the left.
4. The **Probe Cursor** displays the coordinates of the two cursors.

A1=16.000	2.2857
A2=10.000	1.4286
Dif= 6.000	857.143m

In the example shown, **V2** reaches 2.2857 volts when **V1** is at 16.000 volts. **V2** is at 1.4286 volts when **V1** is at 10.000 volts. The voltage differences between the two cursor positions are 6.000 volts for **V1** and 857.143 millivolts for **V2**.

If two or more traces are displayed, the two cursors can be used to display two coordinate points on the same trace or points on different traces. In each case, the difference between the two cursor positions is indicated. For multiple traces, when the cursors are activated, the symbol for the first trace is enclosed within a broken line square. Both cursors will be active on that trace. By right clicking the mouse on another trace, cursor **A2** will be active on that trace. In this way, differences between traces can be obtained.

A Current Source Applied to a Series Circuit

What's new?
1. Applying a current source **IDC** to the circuit
2. The **Plot** command
3. Viewing the **Output file**

Current sources are becoming ever more popular in modern electronics since many semiconductor devices can be modeled as current sources. We shall analyze the circuit shown in Figure 1.03. The circuit is constructed in the usual manner. The **VPLOT** devices are new. They are found in the **Special** library and placed using the same procedure as for the **VPRINT** devices.

Figure 1.03

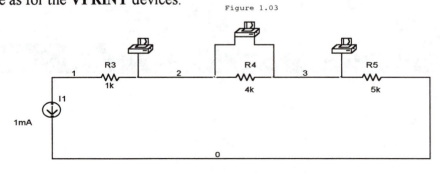

Activating the VPLOT devices
1. Double click on **VPLOT1** until the **OrCad Capture- [Property Editor]** page opens.
2. Scroll to right until **DC** box appears. Click on it and type **Y** in it.
3. Click on **Apply.**
4. Repeat the procedure for the other **VPLOT1** and **VPLOT2** devices.

Viewing the Analysis Results

DC Sweep
The result of a **DC Sweep** analysis produced the **Output file,** of which an edited portion is shown next.

```
*Analysis directives:
.DC LIN I_I1 1mA 1mA 1mA
.PROBE
*Alias File:
**** INCLUDING fig103-SCHEMATIC1.net ****
* source FIG103
R_R3       1 2 1k
R_R4       2 3 4k
R_R5       3 0 5k
I_I1       1 0 DC 1mA
.PLOT      DC V([2])
.PLOT      DC V([3])
.PLOT      DC V([2],[3])
```

```
 ****   DC TRANSFER CURVES           TEMPERATURE =  27.000 DEG C
 LEGEND:
 *: V(2)
  I_I1     V(2)
 (*)---------- -1.5000E+01 -1.0000E+01 -5.0000E+00  0.0000E+00  5.0000E+00

              _ _ _ _ _ _ _ _ _ _ _ _ _ _ _ _ _ _ _
  1.000E-03 -9.000E+00 .        .  *      .        .     . - ------------------------------------
----------------------------
 *: V(3)
  I_I1     V(3)
 (*)---------- -8.0000E+00 -6.0000E+00 -4.0000E+00 -2.0000E+00  0.0000E+00

              _ _ _ _ _ _ _ _ _ _ _ _ _ _ _ _ _
  1.000E-03 -5.000E+00 .        .   *     .        .        .
              - - - - - - - - - - - - - - - - - - - - - - - -

  I_I1     V(2,3)
 (*)---------- -6.0000E+00 -4.0000E+00 -2.0000E+00  0.0000E+00  2.0000E+00
```

```
 ———  —————————————————————————
1.000E-03 -4.000E+00 .          *         .         .         .
 - - - - - - - - - - - - - - - - - - - - - - - - - - -
```

The data shows that a **DC Sweep** analysis was carried out with the current source **I1** held at 1 milliamp. It lists the resistors, their nodal connections and the **PLOT** commands for the indicated voltages. Turning to the plots, the voltage V(3) is listed first with a value of –5 volts. That value is printed out. In addition, there is a linear plot with a range starting at 0.0000E+00 volts and declining to –8.0000E+00 volts. In addition to the numerical value, an asterisk indicates the value of V(3). The voltages V(2) and V(2,3) are shown in the same fashion. With each voltage, the circuit current of 1 milliamp is listed. It is determined by **I1** and is not dependent upon any external elements connected to it. This author is not too enchanted with that kind of graph. It is shown to the reader for informative purposes only.

These voltages are negative because **I1** causes a counterclockwise current flow in the circuit. It enters at the negative terminal of each resistor and leaves at their positive terminal. Since by definition a current leaving a positive terminal is defined as negative by the **PSpice** program, it follows that the product of a negative resistor voltage multiplied by a negative current results in positive power. Thus, it is demonstrated that the resistors consume electrical power. This is as it should be.

Transient Analysis

A **transient analysis** with a ten second duration was performed. The resulting **PROBE** plots show that the voltage traces shown are consistent with the **DC Sweep** data. The voltage of the current source is determined by the external elements connected to it, in our case the three resistors. Application of Kirchhoff's voltage law for the closed circuit shows that the voltage across **I1** is equal to 10 volts.

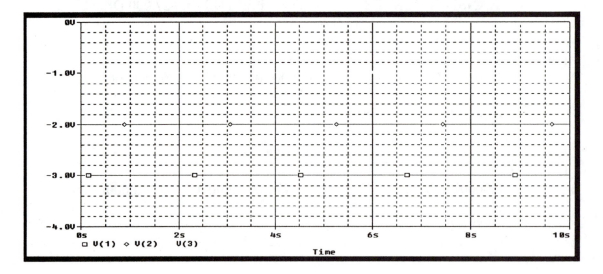

PSPICE ANALYSIS OF A PARALLEL CIRCUIT

Kirchhoff's Current Law for Parallel Circuits

The voltage across each element in a parallel circuit is the same. The algebraic sum of the currents for such a circuit is zero amps. Alternatively, the sum of all the currents through all circuit elements is equal to the source current. The current through each resistor is inversely proportional to its resistance. Thus, the smallest resistor carries the largest current, whereas the largest resistor carries the smallest current. The reader is cautioned to make sure that in the placement of the resistors, their positive terminal is the top one. If this is not done, the currents through the resistors will be negative. The equivalent conductance/resistance of a parallel circuit will be defined.

We shall analyze the circuit in Figure 1.04. All procedures used in the construction of the **Schematic** and in the analysis of the circuit are identical to those of the previous sections.

Viewing the Results of the Analysis: Transient Analysis

The results of a 10 second **transient analysis** are shown on the following **PROBE** plot. The currents through resistors **R1**, **R2** and **R3** are 72.7 milliamps, 20 milliamps and 7.3 milliamps respectively. Their sum, 100 milliamps is numerically equal but of opposite sign to that of the current through **V1**. Thus, as required by Kirchhoff's law for parallel circuits, the algebraic sum of the currents is equal to zero milliamps. As predicted, the smallest resistor, **R1,** with a resistance of 330 Ω carried the largest current, 72.7 milliamps, whereas the largest resistor, **R3,** carried the smallest current of 7.3 milliamps.

Figure 1.04

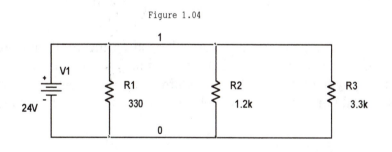

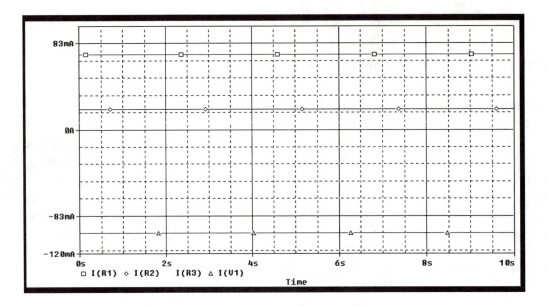

Equivalent Conductance/Resistance of a Parallel Circuit

The current delivered by **V1** was:

$$I(V1) = I(R1) + I(R2) + I(R3) \text{ milliamps}$$

In terms of the circuit voltage, this equation can be rewritten :

$$I(V1) = \frac{V(1)}{R1} + \frac{V(1)}{R2} + \frac{V(1)}{R3} \text{ milliamps}$$

Extracting V(1) as a common factor, we get

$$I(V1) = V(1)\left(\frac{1}{R1} + \frac{1}{R2} + \frac{1}{R3} \right) \text{ milliamps}$$

The quantity within the parenthesis is defined as the **total conductance** of the parallel circuit, symbolized by the letter **G**. Its dimension is the reciprocal of resistance. Its unit is the **siemens**, symbolized by **S**. The determination of the total conductance of a circuit assumes special importance in many electronic circuits. For our circuit,

$$G_{total} = G1 + G2 + G3 \text{ siemens}$$

For our circuit, G_{total} is 4.17 mS.

Its reciprocal is the total resistance of the circuit. Its value is 240 Ω. That resistance is seen by the voltage source V(1). If the three resistors of Figure 1.04 were replaced by a single 240 Ω resistor, the current drawn from V(1) would be the same as that by the original circuit. The proof of this is left, as you may guess, to the reader.

A Series-Parallel Circuit

What's new?
1. A series-parallel circuit with multiple sources
2. Verification of Kirchhoff's current and voltage law
3. Bias Point analysis

As the complexity of circuits increases, the benefits of **PSpice** become quickly apparent. Circuits with multiple sources often require advanced analytic techniques such as superposition to yield a solution. With **PSpice**, increasing circuit complexity simply results in a more complex **Schematic**. Let us analyze the circuit in Figure 1.05.

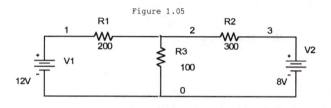

Figure 1.05

We shall run a **Bias Point** analysis to obtain all the absolute node voltages, the source current(s) and the **TOTAL POWER DISSIPATION** of the circuit. After the completion of the construction of the circuit schematic, click on **Simulation Profile** and select **Bias Point** and **Save Bias Point** in the **Simulation Settings** dialog box . The results of the analysis are printed in the **Output File**, an excerpt of which is shown here.

SMALL SIGNAL BIAS SOLUTION   TEMPERATURE = 27.000 DEG C

NODE	VOLTAGE	NODE	VOLTAGE	NODE	VOLTAGE
(1)	12.0000	(2)	4.7273	(3)	8.0000

VOLTAGE SOURCE CURRENTS
NAME   CURRENT
V_V1   -3.636E-02
V_V3   -1.091E-02

TOTAL POWER DISSIPATION 5.24E-01 WATTS

V(1) and V(3) are the nodal voltages of **V1** and **V3**. Node 2 is at 4.7273 volts. Since this is less than the voltage of either node 1 or node 3, both source currents flow into Node 2. Their sum, 47.27 milliamps, flows through **R3** and causes a voltage drop of 4.7273 volts. This confirms that the net sum of the current into node 2 is zero amps. Kirchhoff's current law is confirmed.

If we sum the voltages around any closed path in the circuit, their algebraic sum should be equal to zero volts. Let us start at the 0 node, go upward through **V1**, next from node 1 to node 2, and from node 2 back to node 0.

$$\text{Thus: } V(0) + V(0,1) + V(1,2) + V(2) = 0 \text{ volts}$$
$$\text{Substituting : } 0V + (-12V) + (12 - 4.742)V + 4.72 = 0 \text{ volts}$$

This confirms **Kirchhoff's voltage law**. The reader is encouraged to try any other path to confirm that law. The analysis was run at 27 DEG Centigrade. This corresponds to 80.6 DEG Fahrenheit. We will learn that we can change this temperature and affect the behavior of a circuit.

Dependent Sources

A Voltage-controlled voltage source

What's new?
1. The use of a voltage-controlled voltage source
2. Rotation and mirror imaging of a dependent source

The voltage and current sources used so far had terminal voltages and currents which were independent of any circuit element(s) connected to them. Such sources are defined as independent sources. There are voltage and current sources whose voltages and currents depend upon some other voltage or current elsewhere in a circuit. Such sources are defined as dependent sources. There are four kinds:

Name of source	**PSPICE** symbol
Current-controlled current source(CCCS)	F
Current-controlled voltage source (CCVS)	H
Voltage-controlled current source(VCVS)	G
Voltage-controlled voltage source(VCVS)	E

PSpice allows for considerable savings in time and effort in the analysis of circuits containing such sources compared to traditional analytic methods. Also, many solid state devices can be modeled by means of dependent sources. Thus, the modeling of dependent sources is an important asset in the analysis of circuits containing such devices. We begin our analysis with the circuit shown in Figure 1.06.

It contains a voltage-controlled voltage source. Note the letter **E** next to the symbol of the source. The voltage at node 2 is the controlling voltage. The controlling terminals are connected across **R4**, much like a voltmeter would be connected. The voltage at node 3 is the controlled voltage. The forward voltage gain of the **E** source is defined as the dimensionless ratio of the voltage across the controlled terminals divided by the voltage across the controlling terminals. In this circuit, the **E** source has a voltage gain of

100. Since the controlling voltage is 20 millivolts, we should obtain 2 volts for the voltage at node 3.

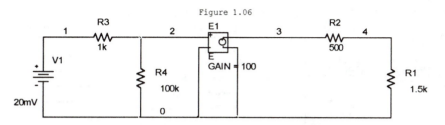

Figure 1.06

To obtain the **E** source:
1. In **Place Part** dialog box, scroll to and click on **ANALOG** in the **Libraries** box.
2. Click in the Part box, type E or scroll to E and click to select.
3. Click on **OK;** the **E** symbol will appear on screen.
4. Move to the desired location and click to place.
5. Click right; click left on **End Mode** to end placement.
6. Complete the circuit per previous procedures.

Rotation and mirror image of E source
      The **E** source can be rotated in the same way that resistors are rotated. In addition, should it be desired to obtain the mirror image of the E source, proceed as follows:
1. Click on the **E** source to select it.
2. From **Edit** menu, click on **Mirror**. The following options appear:

HORIZONTALLY H
VERTICALLY V
BOTH

Select to choose.

      We shall run a **Bias Point** analysis to obtain all the node voltages, the source current and the TOTAL POWER DISSIPATION of the circuit. At the completion of the construction of the circuit schematic, click on **Simulation Profile,** select **Bias Point** and **Save Bias Point** in the **Simulation Settings** dialog box, as shown.

```
Simulation Settings - 2                                              ⊠
 General   Analysis │ Include Files │ Libraries │ Stimulus │ Options │ Data Collection │ Probe Window │

 Analysis type:              ┌ Save bias information in filename:
 │Bias Point        ▼│       │                                         │    Browse... │

 Options:                    ┌─ Options ──────────────────────────────────────────────
 ☑General Settings             Save bias information:
 ☐Temperature (Sweep)            When Primary Sweep value is:          │          │
 ■Save Bias Point                When Secondary Sweep value is:        │          │
 ☐Load Bias Point                When Parametric Sweep value is:       │          │
                                 When Monte Carlo run number is:       │          │
                                 When Temperature Sweep temperature is:│          │

                               ☐ Do not save subcircuit voltages and currents

              OK   │   Cancel   │   Apply   │   Help
```

Click on **OK**. Run the analysis. It produced the data summarized below.

Node voltage 1	.02 volts
Node voltage 2	.019 volts
Node voltage 3	1.98 volts
Node voltage 4	1.49 volts
Source current	-.20 µamps
TOTAL POWER DISSIPATION	.004 µwatts

From the data we note that V(2) is .019 volts and V(3) is 100 times that voltage because of the gain of the **E** source. The voltage across **R2** is

$$V(3)-V(4) = 1.98 \text{ V} - 1.49 \text{ V} = .49 \text{ volts}$$

This makes the current through **R2** equal to .98 milliamps. This current is far larger than that delivered by **V1**. This increase in current is due to the voltage gain of the **E** source. Such an increase in current is typical for amplifier circuits containing bipolar transistors (BJT). The **E** source could be the model for such a device.

The TOTAL POWER DISSIPATION of .004 µW listed here is that of **V1**. Multiplying The VOLTAGE SOURCE CURRENT of - 20µA by the .02 V of **V1** equals the power delivered by **V1**. A positive TOTAL POWER DISSIPATION means that **V1** is losing power. A negative TOTAL POWER DISSIPATION would mean that **V1** is gaining power.

A Current-controlled current source

What's new?
1. The use of a current-controlled current source
2. The use of **WATCH1**
3. The use of multiple Y-axes

We shall analyze the circuit shown in Figure 1.07. The current through **R2** or **R3** controls the current that flows out of the **F** source and into node zero. The controlled and the controlling terminals of the **F** source are connected like an ammeter into the circuit. The forward current gain of the **F** source is defined as the dimensionless ratio of the current flowing out of the controlled terminals divided by the current though the controlling terminals. In our case, let this ratio be 100.

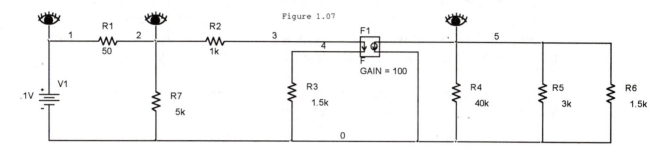

Figure 1.07

The gain of the source of **F** is set by double clicking on its symbol. The **OrCAD Capture [Property Editor]** dialog box shown opens. Scroll to **GAIN** box and click in it. Type 100. This will set the gain of the **F** source. Click on **Display** and the **Display Properties** dialog box opens. Click on **Name and Value.** Click on **OK**; the **Display Properties** box closes. In **Properties Editor** box, click on **Apply**. Click on **Close(X)** in the upper-right-hand corner to exit **Properties Editor**. The legend GAIN = 100 will appear next to source **F1** on the schematic screen.

The new symbol introduced in Figure 1.07, **WATCH1,** allows the print-out of up to three circuit voltages in a **DC sweep**, **TRANSIENT** or an **AC sweep** analysis. The latter will be covered in future chapters. To place **WATCH1**, in **Place Part** dialog box, select **SPECIAL** from **Library**. Click on **WATCH1** to select, click on **OK**. Place part at desired location(s).

To activate **WATCH1**, double click on its symbol. The **OrCAD Capture- [Property Editor]** dialog box opens. Scroll to **Analysis** box. Type analysis desired, in our case, type **TRAN**. Scroll to **LO** and **HI** boxes. Enter estimated values of the voltages to be recorded. The author used 0 V for **LO** and –5 V for **HI**. Repeat for the other two devices. Click to close in each case. Run the **TRANSIENT** analysis.

We shall perform a **transient analysis** of 10 seconds duration. The analysis printed out the results shown in the **Output File.**

Node 1	.1000 volts
Node 2	.0971 volts
Node 3	.0583 volts
Node 4	.0538 volts
Node 5	-3.7888 volts

Node 1 displays the source voltage **V1**. Its value is .1 volts. The voltage V(2) is close to that voltage because of the voltage divider formed by the 5 kΩ and the 50 Ω resistors. The voltage V(3) is further diminished because of the voltage divider formed by **R2** and **R3**. The voltage V(4) is identical to V(3) because there is no voltage drop across the controlling terminals of the **F** source.   Of special interest is the voltage V(5). Its voltage of -3.7888 volts is determined by the current through the **F** source and the circuit elements connected to it.   Should the resistance of **R4**, **R5** or **R6** change, the controlled current through the **F** source would not change. The voltage across its controlled terminals would change.   The negative sign of V(5) is because the current flows downward from the **F** source into node 0. This forces the currents in resistors **R4**, **R5** and **R6** to flow from node zero toward node 5.

We can display both the circuit voltages and currents on one **PROBE** plot by using multiple Y-axes. To obtain the second Y-axis:
1. Click on **Plot.**
2. Click on **Add Y-axis**. It will appear on the **PROBE** plot. The symbol >> next to a Y- axis denotes the active one. To change the active Y-axis, click on the other one. The symbol >> will move to the other Y-axis.

3. Click on **Trace** to select the desired variable. The **PROBE** plot shown has two Y-axes. The current axis is labeled as 1 and the voltage axis as 2.

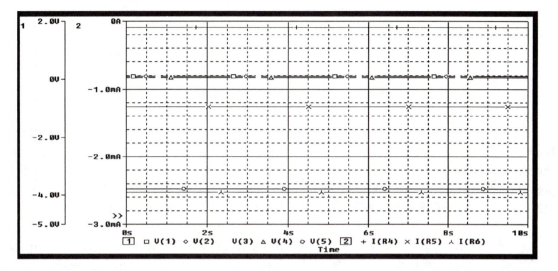

The node voltages shown have the same values as in the **Output File**. The sum of the currents through resistors **R4**, **R5** and **R6** are equal to the current through the controlled terminals of **F**. Its value is −3.8835 milliamps as determined by cursor **A1** placed at t = 4 seconds. The current through the controlling terminals is listed in the **Output File** and it is given as 3.884E-05 amps. Thus the ratio of the controlled over the controlling current is 100. This is equal to the current gain of the **F** source.

Loading Effects of Voltmeters in Circuits

A voltmeter measures voltages across a circuit element such as a resistor. By placing a voltmeter across a resistor, a parallel circuit is created in which the resistance of the resistor is in parallel with the input resistance or input impedance of the voltmeter.

Ideally, the input impedance of the voltmeter should be infinite. This would insure that measurement of the voltage across a resistor would not effectively change the circuit. In practice, this ideal can only be approached. Typically, volt-ohm-ammeters, or VOMs, have an ohm/volt rating of 20,000 Ω/V. If the 10 V scale of such a meter were used to measure a voltage, a resistance of 200 kΩ would be placed in parallel with the element across which the voltage is to be measured.

By contrast, digital multimeters, or DMMs, have an input impedance of about 10 MΩ regardless of which of its volt scales is used. We shall use **PSpice** to obtain the effect of the two types of meters on circuits with different resistance values. There are no new **PSpice** devices, commands or procedures needed for the analysis to follow.

Low-Resistance Measurement with VOM and DMM

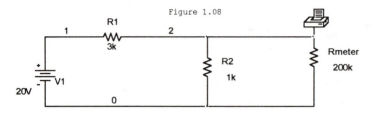

Figure 1.08

For the circuit in Figure 1.08, it is required to measure the voltage across **R2** with both a VOM and a DMM. To select the correct scale for the VOM, we compute that the theoretical voltage across **R2** is equal to 5 V because of the voltage divider formed by **R1** and **R2**. Selecting the 10V scale of the VOM places 200 kΩ in parallel with the 1 kΩ of **R2**. Running a **DC Sweep** analysis gives 4.98 V as the voltage across **R2**.

We next replace the VOM with a DMM. Regardless of which of its voltage scales are used, its input impedance is 10 MΩ. This resistance is now parallel with the 1 kΩ resistance of **R2**. Running a DC Sweep analysis yields 5 volts for the voltage across **R2**.

High-Resistance Measurements with VOM and DMM

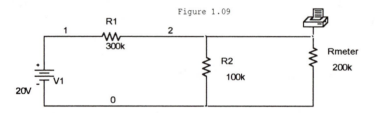

Figure 1.09

We shall next measure the voltage across **R2** in Figure 1.09. Again, its theoretical voltage is 5 V. We select the 10 V scale on the VOM. Its 200 kΩ input impedance, or resistance, is in parallel with the 100 kΩ of **R2**. Running a **DC Sweep** analysis, we obtain 3.64 V for the voltage across **R2**.

We next replace the VOM with the DMM. Its input impedance of 10 MΩ is now in parallel with the 100 kΩ of **R2**. Running a **DC sweep** analysis yields 4.96 volts for the voltage across **R2**.

The analysis results are summarized in the following table:

Meter used	R2	V(2) ideal	V(2) measured	% change
VOM	1 kΩ	5 V	4.98 V	-.4
DMM	1 kΩ	5 V	5.00 V	0.0
VOM	100 kΩ	5 V	3.64 V	-27.2
DMM	100 kΩ	5 V	4.96 V	-.8

From this data, it is apparent that when both meters were used to measure the voltage across **R2** in Figure 1.08, the measurement error was negligible for theVOM and nonexistent for the DMM. However, in Figure 1.09, the use of the VOM, when placed in parallel with the 100 kΩ of **R2** resulted in a 27.2 % reduction of V(2). Such is clearly unacceptable. By contrast, the measurement error introduced by the DMM was still only a .8 % reduction.

Electrical power circuits generally have low resistance values. Hence, the use of a VOM will result in small measurement errors. However, electronic circuits often have high resistance values; hence a VOM cannot be used without introducing unacceptable measurement errors. In such circuits, the use of a DMM with its high input impedance is mandated.

PARAMETRIC ANALYSIS

What's new?
1. The use of a variable resistor **Rvar** of values **Rval**.
2. **Parametric Sweep**
3. Selective **PROBE** plots

Figure 1.10 is a voltage divider. It is often used to control the output voltage of an electrical system. The volume control of a radio or a television set is an example that comes to mind.

Figure 1.10 has a variable resistor, labeled **Rval**. We shall require that it changes its resistance from 100 Ω to 2100 Ω in steps of 400 Ω. The **PSpice** program will perform iterative calculations for each of the stipulated values of **Rval**. For the range of **Rval** and its increments, six runs will be performed by **PSpice**. At the completion of the analysis, **PROBE** allows for the viewing of all the traces of a selected variables such as V(2) corresponding to each resistance value of **Rval**. Also, we can specify the value of V(2) to be plotted for a particular run of the program.

Constructing the circuit
1. Get the voltage source **V1** and place it. Set its voltage to 10 V. Get the first resistor **R1** and place it.
2. Get the second resistor **R2** and place it.
3. Wire the circuit and label nodes.
4. Double click on the 1 kΩ resistance of **R2**. The **Display Properties** box opens.
5. In it, replace 1k with **{Rval}**. The braces are a must! Click on **OK**. **R2** now has **Rval** as its resistance on Schematic screen.
6. From **SPECIAL** library, scroll to and select **PARAM**. Click on **OK** and place it in open area on the screen.
7. Double click on it, the **Property Editor** sheet opens. Click on **New**.

8. In the **Property Name:** box, type **Rval** without the braces. Click on **OK**. A new box, labeled **Rval,** appears in the **Property Editor** sheet.
9. In that column, enter initial value for **Rval,** 100 Ω.
10. Click on **Display**
11. In **Display Format** box, select **Name and Value**. Click on **OK**.
12. Click on **Apply** to verify all data entries made. Close **Property Editor** sheet.

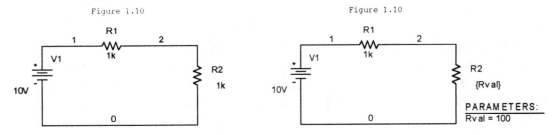

After completion of step 12, Figure 1.10 will appears as shown on right above.

Running the analysis

We shall run **parametric analysis** together with a **transient analysis** of 10 second duration. Its parameters are entered as shown.

1. Click on **PSpice,** click on **New Simulation Profile.**
2. In its **Name** box, type parametric. Click on **Create.** The **Simulation Setting** dialog box opens.
3. Enter **transient** parameters as shown.

Simulation Settings - parametric2

General   Analysis   Include Files   Libraries   Stimulus   Options   Data Collection   Probe Window

Analysis type:

Time Domain (Transient) ▾

Options:

☑ General Settings
☐ Monte Carlo/Worst Case
☐ Parametric Sweep
☐ Temperature (Sweep)
☐ Save Bias Point
☐ Load Bias Point

Run to time: 10s   seconds (TSTOP)

Start saving data after: 0   seconds

Transient options

Maximum step size: 1s   seconds

☐ Skip the initial transient bias point calculation (SKIPBP)

Output File Options...

OK   Cancel   Apply   Help

4. Click on Parametric Sweep
5. Enter sweep parameters as shown

6. Click on **OK**. The **Schematic**s screen opens up again.
7. Click on **PSpice**.
8. Click on **Run**. The **Available Sections** box opens.
9. Click on **All** to see all traces in the PROBE plot.
10. Click on **OK**. The PROBE screen opens.
11. Select trace of V(2) to see all values of V(2) for each value of Rval.

Viewing the Analysis Results
The **PROBE** plot shows the voltage of V(2) for each value of **Rval.**

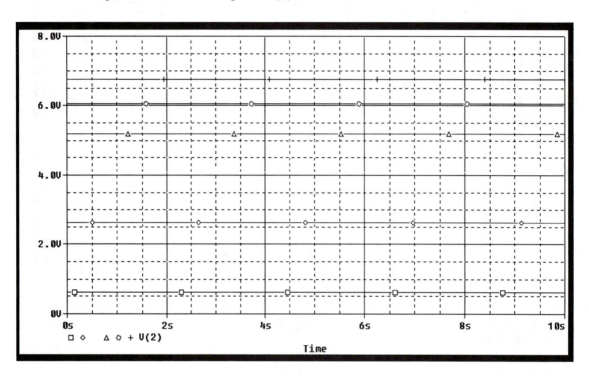

We have the option to select among the traces we want to have plotted. For instance, should we desire to see only the third run of the analysis, we would type V(2)@3. Only its trace would appear on the **PROBE** plot. To compare the first with the last run of the program, we would type V(2)@1 V(2)@6. Only those two traces would be plotted.

The Output file
For ease of reference, the values for V(2) printed in the **Output File** have been listed next.

Rval Ω	V(2) volts
100	.90
500	3.33
900	4.73
1300	5.65
1700	6.29
2100	6.77

PROBLEMS

1.1   The circuit shown consists of five series resistors. Find the voltages across all
        resistors and the circuit current.

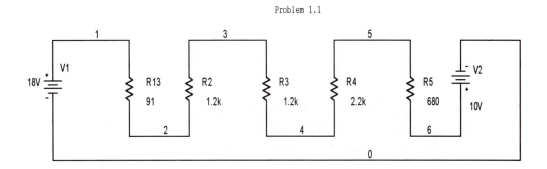

Problem 1.1

1.2   The circuit shown resembles that of a dc armature. Find and print out the voltages
        across all resistors and the voltages across the node pairs 8-2 and 5-11. Find and
        print out the armature current. All resistors have a resistance of .5 Ω.

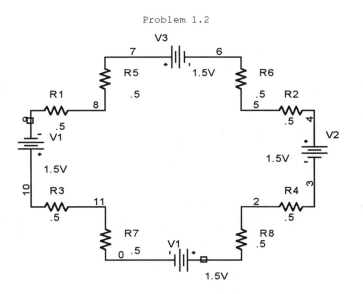

Problem 1.2

1.3   A Simpson 260 voltmeter has voltage scales of 10 V, 50 V and 100 V.
        The voltmeter has a current sensitivity of 50 µA or 20 kΩ/V.
        The 10 V scale is to be used to measure the voltage across **R2**. Calculate
        the impedance of the voltmeter. Calculate the theoretical value
        of that voltage. Perform a **PSpice** analysis to obtain the voltage

across **R2**. Compare the measured voltage with its theoretical value.
Compute the % change between these two, using the theoretical voltage
as the standard of comparison.

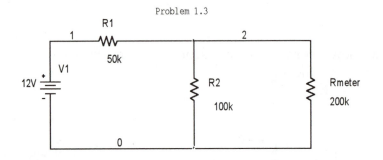

Problem 1.3

Use 10V scale to measure voltage across R2

1.4   The voltmeter and the circuit of Problem 1.3 are used. However, the 50 V scale on
the Simpson meter is selected to obtain the voltage across **R2**. Calculate the new
value of the impedance of the meter. Perform a **PSpice** analysis to obtain the
voltage across **R2**. Compare the new reading with the one obtained before. Which
of these two readings is nearer to the theoretical voltage across **R2** and why?
Which of the two readings reflects a larger "loading effect" of the Simpson 260 on
the circuit?

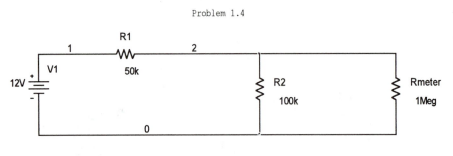

Problem 1.4

Use the 50 V scale on the meter.

1.5   The circuit shown consists of two fixed voltage sources **V1** and **V2**, their associated
resistors and a load resistor, all connected in series. Use the **PSpice** program to
print all the absolute and the relative voltages and the current flowing in the circuit.

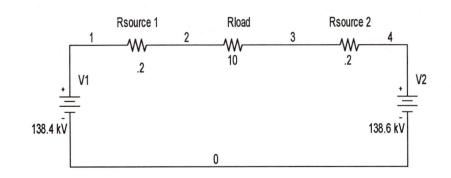

Problem 1.5

1.6   This problems uses the same circuit as Problem 1.5. However, voltage source **V2** changes its voltage from 138.6 kV to 138.2 kV in steps of 20 V. Use **PROBE** to print and plot all voltages and the current resulting from that voltage change. At what voltage of **V2** does the current change direction? How does the power flow of the two voltage sources change as the voltage of **V2** changes? Does the direction of the power flow to the resistors change?

1.7   **V1** is a 9 V dc voltage source and **I1** is a current source. It can change from 1 A to 2 A in steps of .1 A. Resistor **R1** has a power rating of 20 watts. Find the maximum allowable current in this circuit based on the power rating of the resistor. Also, for that current, what is the power delivered to **R1** by **V1** and by **I1**?

Problem 1.7

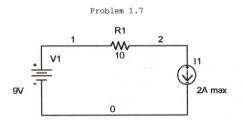

1.8   A 1200 Ω resistor with a 1 W power rating and a 470 Ω resistor with a rating of .25 watts are connected in series to a variable dc voltage source. To ensure that the power rating of the 470 Ω resistor is not exceeded, its maximum voltage is to be kept at or below 10volts. Find the maximum permissible voltage of the input voltage source **Vin.**

Problem 1.8

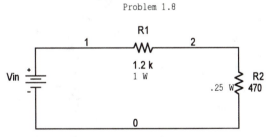

1.9   A d'Arsonval meter movement with a 1000 Ω/V sensitivity has a multiplier resistor of 9 kΩ connected to it in series. Write a **PSpice** program to determine the maximum voltage that can be measured with this meter.

1.10   Voltage sources **V1** and **V2** and resistor **Rload** are connected in series. **V1** changes its voltage from 5 volts to 20 volts in steps of 1 volt.

Problem 1.10

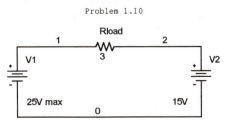

Using **PROBE**, obtain the following:
(a) What is the voltage across **Rload** for **V1** = 5 volts, 15 volts and
     20 volts respectively?
(b) Compare the amplitude and polarity of the source currents with the amplitude
     and polarity of the current through **Rload.**
(c) What does the polarity of the source currents indicate?
(d) Obtain the power to **Rload** for **V1** = 10 volts, 15 volts and 20 volts.
(e) Is the power to the resistor always positive?
(f) When is that power a maximum and when a minimum value?
(g) Obtain the power flow of **V1** and **V2** for **V1** = 10 volts, 15 volts and 20 volts.
(h) Below V1 = 15 volts, which voltage source delivers power, which receives
     power?
(i) At **V1** = 15 volts, what is the power flow for **V1, V2** and **Rload**
(j) Above **V1** = 15 volts, which voltage source receives power,
     which delivers power?

1.11   For the circuit shown, obtain the following:
       (a) The current through each resistor
       (b) The total current into the circuit
       (c) The total conductance of the circuit
       (d) The total resistance of the circuit
       (e) The power to each resistor
       (f) The total power delivered by the voltage source
       (g) Perform a power audit that compares the power used by the resistors
            with the power delivered by the voltage source.
       (h) Which resistor consumes the smallest amount of power?
       (i) Which resistor consumes the largest amount of power?

Problem 1.11

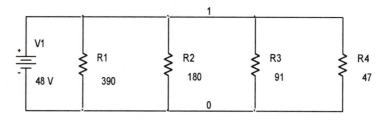

1.12   For the circuit shown, do the following:
       (a) Find the current through each resistor.
       (b) Explain the sign convention of the currents.
       (c) Verify Kirchhoff's current law for each node.
       (d) Find the circuit voltage and explain its polarity.
       (e) Find the circuit resistance as seen by the current source.
       (f) What name is assigned to that resistance?

(g) Find the power delivered to each resistor and compare it to the
power delivered by the current source **I1**.

Problem 1.12

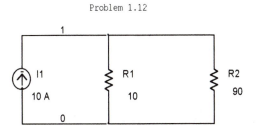

1.13   The current source **I1** can change its current from 50 milliamps to 300 milliamps
in steps of 10 milliamps. All resistors are rated at 2 watts of power. Find the
maximum allowable current of **I2**. Hint: The smallest resistor will
receive maximum power. Make sure that no resistor receives more
than its allowable power. At the maximum allowable current of **I2**,
what is the voltage across the resistors?

Problem 1.13

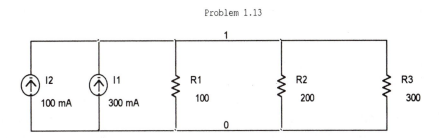

1.14   In the circuit shown, the voltage source **V1** can vary between zero volts and
50 volts in steps of 1 volt. Given the .5 watt rating of the resistors, find the
maximum permissible voltage of **V1**. Which resistor limits **V1** to that value?

Problem 1.14

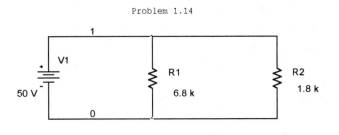

The resistors are rated at .5 W each.

1.15   Given the circuit shown:
(a) Find the current in each resistor.
(b) Find the power to each resistor.
(c) Verify Kirchhoff's current law.
(d) Resistor **R3** develops an open circuit across it. Find all the
circuit currents and the power to each resistor.

(e) You have just fixed **R3** when **R1** becomes short-circuited. Find
all the resultant currents and the power to each resistor.

Problem 1.15

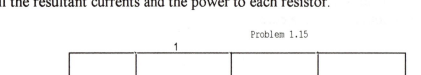

1.16   Design a heater circuit that is to deliver a total of 1200 watts of electrical power.
The design is to consist of the correct number of 200 Ω resistors, each rated at 200
watts. Find the needed voltage source. Select a proper fuse that limits the total
current to a safe value. Use **PSpice** to implement and check your design.

Problem 1.16

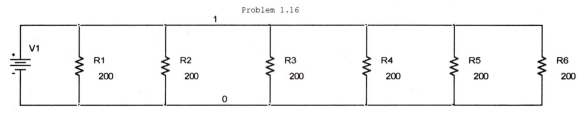

All reistors are rated at 200 watts.

1.17   **I2** is a current source that can change its current from 1 milliamp to 40 milliamps
in steps of 1 milliamp. **I1** is a fixed current source with a current of 20 milliamps.
(a) Find the value of the resistor currents for the range of **I2**.
(b) What is the magnitude of the resistor currents when **I2** = 20 milliamps?
(c) How is the polarity of the resistor current effected over the
range of the current of **I2**?
(d) What is the minimum power to the resistors?
(e) At what amplitude of the current of **I2** does it occur?
(f) At what values of **I2** is the power to **R2** equal to 400 milliwatts?
(g) When the current through **I2** is 20 milliamps, what is the power delivered or
absorbed by **I1** and **I2**?
(h) What is the power flow for the two current sources below?
I2 = 20 milliamps?
(i)   What is the power flow for the two current sources above?
I2 = 20 milliamps?

Problem 1.17

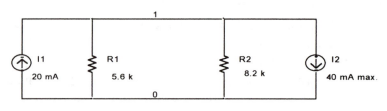

1.18   A student-designed **VOM** has a sensitivity of 1000 Ω/V and three scales of 15 V, 150 V and 1000 V. It is used to measure the voltage across the 180 kΩ resistor. Use **PSpice** to obtain the measurement error expressed as a percent deviation introduced by this meter.

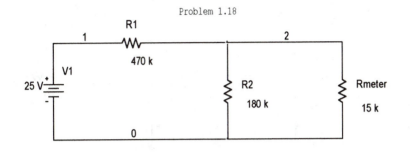

The 15 V scale is used to measure the voltage across R2.

1.19   For the circuit shown, find all nodal voltages and the current through each resistor. For a closed loop within this circuit, verify Kirchhoff's voltage law. For a node of the reader's choice, verify Kirchhoff's current law. The output data can be obtained in both the **Output file** and on a **PROBE** plot.

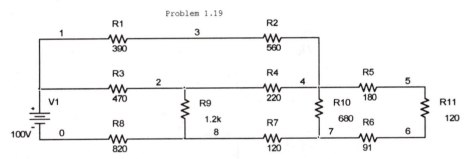

1.20   For the current-divider circuit shown, find all nodal voltages with respect to ground. Obtain the current through each resistor. This can be obtained from its **PROBE** trace. Determine the total resistance as seen by the current source **I1**.

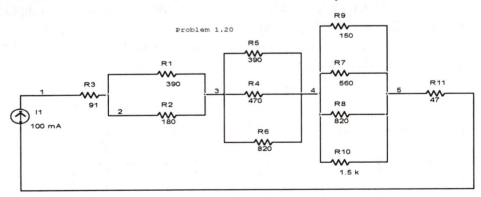

1.21   The circuit resembles that of a dc armature. Find the voltages at all nodes and the current through each resistor. From these values, compute the total power into the circuit delivered by the four batteries.

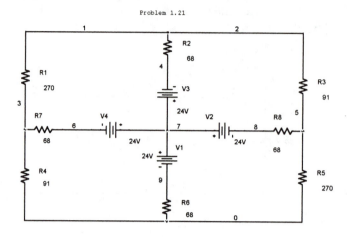

Problem 1.21

1.22   The circuit has both an independent voltage and current source. Find the voltage at each node, the voltage across each resistor and the current through each of them. What are the power delivered by the voltage source and that delivered by the current source?

Problem 1.22

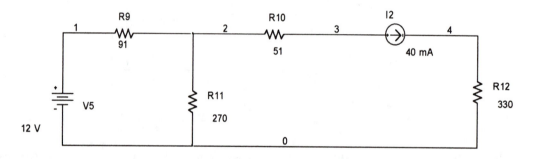

1.23   The circuit has two independent current sources. Find all nodal voltages, the voltage across each resistor and its current. Get the total power consumption of the circuit and the power delivered by each current source. Verify Kirchhoff's voltage law for any closed path in the circuit. Verify Kirchhoff's current law for at least three nodes of your choice.

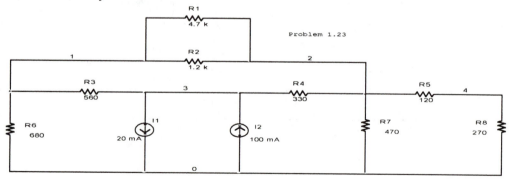

Problem 1.23

1.24   **V1** changes from -20 volts to 20 volts in steps of 1 volt. Obtain the numerical and the graphical output data of the voltage across and the current through **R3**.

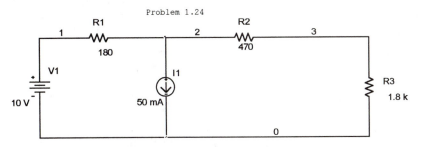

Problem 1.24

1.25   The given circuit contains a fixed voltage source **V1** and a current source **I1** that can change its current from –20 milliamps to 20 milliamps in steps of 2 milliamps. Find all the nodal voltages and plot the voltage across **R4** as a function of the current of **I1**.

Problem 1.25

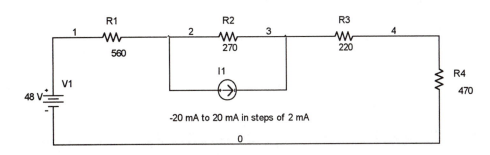

1.26   **V1** is a fixed voltage source of 125 volts. **V2** is a variable voltage source that can change its voltage from 110 volts to 130 volts in steps of 1 volt. Find the permissible range in the voltage of **V2** that keeps the voltage across **R6** between 56 volts and 57 volts. Obtain both graphical and numerical output data of that voltage.

Problem 1.26

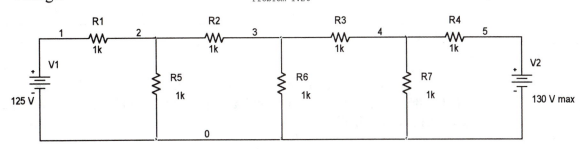

V2 can change from 110 V to 130 V maximum in steps of 1 V.

1.27   The circuit has an independent voltage source **V3**. Its voltage can change from 1
        millivolt to 10 millivolts in steps of 1 millivolt. The circuit also has a dependent
        voltage source **E1**. It has a forward voltage gain of 260. Its controlled terminals are
        connected across **R3** and **R4**. Find the node voltages V(3) and V(4) when the
        source **V3** is 6 millivolts. Find the ratio of V(3)/V(2) at that voltage of **V3**.

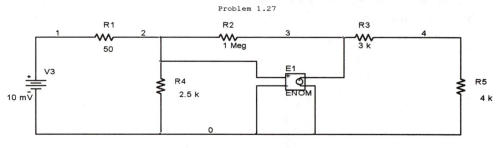

Problem 1.27

E1 has forward voltage gain of 260.

1.28   **V1** changes from 100 millivolts to 200 millivolts in steps of 5 millivolts. A
        dependent voltage-controlled current source **G1** with a transconductance of 20 mS
        develops a current in response to the voltage across **R3**. This circuit resembles an
        amplifier model with the **G** source representing a BJT. Find the following:
        (a) The value of the current through the **G** source when **V1** = 100 millivolts,
             160 millivolts and 200 millivolts.
        (b) Determine the ratio of V(3)/V(1).
        (c) What is the polarity of that ratio and why?

Problem 1.28

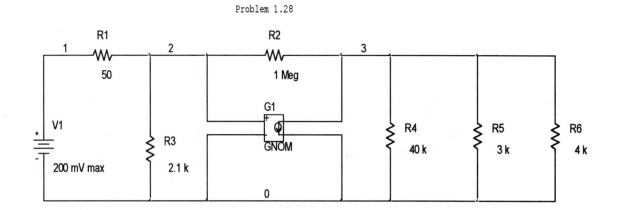

1.29   The circuit contains an independent voltage source **V1** and two dependent voltage-
        controlled voltage sources. Both have voltage gains of 2. Find the nodal voltage V(2)
        and the currents through all resistors.

Problem 1.29

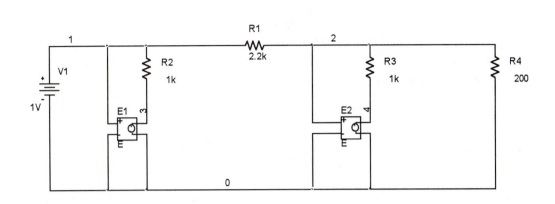

1.30   The circuit has an independent voltage source **V1** and a dependent current-controlled current source **F1**. Its forward current transfer ratio is 125. The controlling current flows through **V1**. Find all nodal voltages, the currents through **F1, R2** and **R3**. Obtain the voltage gain defined as V(2)/V(1,3) of this circuit. Verify that the current gain of **F1** is 125.

Problem 1.30

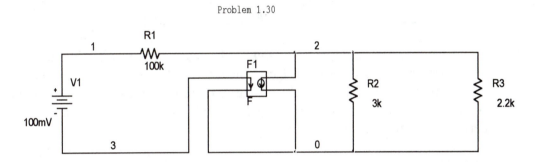

1.31   The circuit shown has two current-controlled voltage sources, **H1** and **H2**. Both are controlled by the current through **V1** and **R1**. **H1** has a transresistance of 100 Ω and **H2** has a transresistance of 50 Ω. Find all absolute and relative voltages, including the voltages across the two dependent sources. Verify the transresistance of the two **H** sources. Explain their polarity. In this problem, a ground symbol has been introduced. To obtain it: Click on **Place**, click on **Ground**, in **Place Ground** box, scroll to desired symbol. Click on **OK**. Move the ground symbol to desired location on screen. **Click** to place. **Click** right to end placement.

Problem 1.31

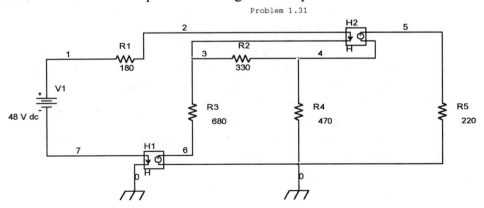

1.32   The circuit has a current-controlled dependent voltage source **H1** with a transresistance of 100 Ω. The controlling current flows through resistor **R6**. Obtain the voltage across the controlled terminals of the **H** source. Verify its transresistance at t = 6 ms.

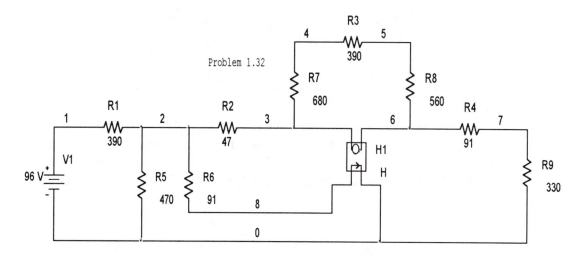

Problem 1.32

1.33   The circuit shown contains a variable voltage source **V1** and a current-controlled dependent current source **F1**. **V1** can change its voltage from 10 millivolts to 100 millivolts in steps of 1 millivolt. **F1** has a current gain of 1000. The controlling current flows through a resistor **R3**. **R5** is rated at 1/2 W. Obtain the maximum permissible value of **V1**.

Problem 1.33

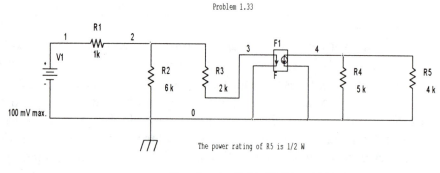

V1 can change from 10 mV to 100 mV in steps of 1 mV.

1.34   The circuit contains two voltage-controlled dependent voltage sources, **E1** and. **E2**. Source **E1** has a voltage gain of 5. It is controlled by the voltage across **R4**. Source **E2** has a voltage gain of .1. It is controlled by the voltage across **R5**. The circuit consists of two loosely coupled loops in which part of the voltage across **R5** is fed back into the loop containing **V1**. Such a circuit is defined as a negative feedback circuit. Obtain the following:
(a) The voltage across **R5** when **V1** = 80 millivolts and 40 millivolts
(b) The controlling voltage for **E1** when **V1** = 80 millivolts and 40 millivolts

(c)   The controlling voltage for **E2** when **V1** = 80 millivolts and 40 millivolts
(d) Verify the voltage gains of the two **E** sources.

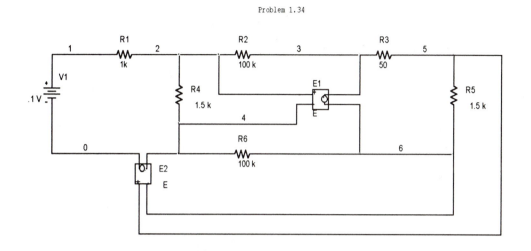

1.35   Set the gain of the **E2** source to zero. Obtain all the voltages as in Problem 1.34.
Compute the percent deviations of the present voltages compared to those in
Problem 1.34. Use these latter voltages as the standard for your comparison. The
deviations you calculated are due to the effect of the voltage feedback caused by
E2.

1.36   For this circuit, **Rval** can change from a minimum of 100 kΩ to 1 MΩ in steps of
100 kΩ. Find the voltages of V(3) for each value of **Rval**. Obtain a **PROBE** plot
that shows the second, the fourth and the sixth run of the program.

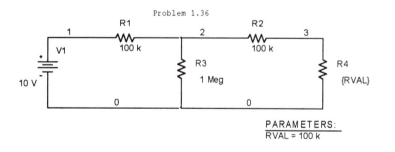

DC NETWORK THEOREMS

THE PRINCIPLE OF LINEARITY

We begin with the definition of linearity. A linear circuit element has a constant relationship between the voltage across it and the current through it. The voltage-current relationship for a resistor, neglecting for the moment its temperature dependence, is:

$$v(t) = Ri(t) \text{ volt}$$

The voltage and the current can be functions of time. Their ratio is a constant. If we plot the ratio of the voltage divided by the current, we will find that the ratio plots as a straight line, hence the name linear. Such a graph is said to have a constant slope. In the case of the above equation, that slope is the resistance **R**. If we increment the current by a certain amount, the voltage will be incremented in such a way that the ratio of voltage to current still remains.

The principle of linearity also holds for dependent sources. For instance, a linear dependent source is a source whose voltage or current is in proportion only to the first power of some independent voltage or current source in a circuit. For example

$$V_{dependent} = 15V1 + 6I1 \text{ volts}$$

is an example of a linear-dependent source. Its voltage depends only on the first powers of **V1** and **I1**. By contrast:

$$V_{dependent} = 15(V1)^2 \text{ volts} \quad \text{and} \quad V_{dependent} = 15V1 * 6I1 \text{ volts}$$

are examples of nonlinear equations since the dependent voltage depends upon the second power of **V1** and the second power of **V1*I1.**

Demonstration of Linearity

We shall use the circuit in Figure 2.01 to demonstrate the **principle of linearity**. The input voltage V1 is to change from 1 volt to 30 volts in steps of 1 volt.

Figure 2.01

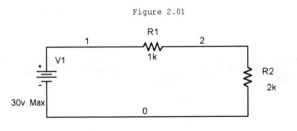

The **PROBE** plot shows the traces of the nodal voltages of V(1) and V(2). Both are linear plots. For any point along the horizontal axis, the ratio of V(2)/V(1) is constant at .667, as determined by the ratio of **R2/(R1+R2).** For any incremental changes in V(1) and consequently in V(2), the ratio of V(2)/V(1) will always be .667. Such is the property of linearity.

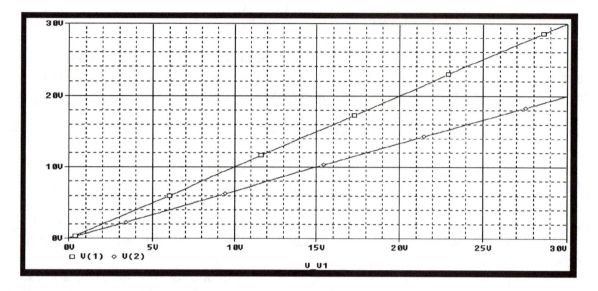

We can get a **PROBE** plot of this ratio by requesting a plot of V(2)/V(1). This was done, and the result is shown.

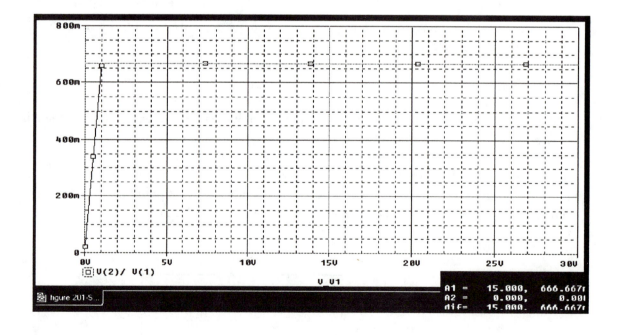

The cursor **A1** placed at V(1) equal to 15 volts shows that the ratio of V(2)/V(1) is equal to 0.666. The point has been made.

PROBE is capable of performing arithmetic and transcendental operations. A listing of these capabilities can be found in the **Functions or Macros** scroll box shown.

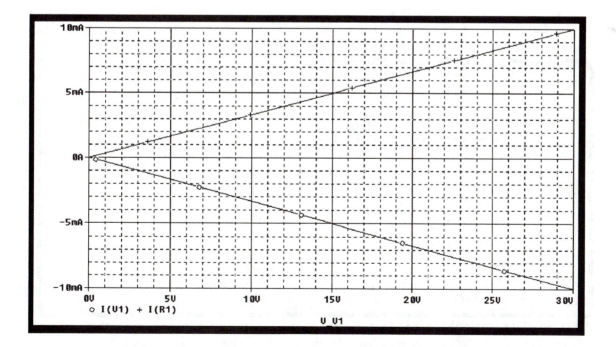

We next plot the current in the circuit.

The plots show the linear variation of the current through **R1** and **V1**. By the syntax of **PSpice**, the current entering the positive terminal of **R1** is defined as a positive current, whereas the current leaving the positive terminal of **V1** is defined as negative. If we let **PROBE** plot the ratio of V(1)/I(R1), we obtain the total resistance of the circuit. The **PROBE** plot shows that the total resistance of the circuit is 3 kΩ. Incidentally, the expression I(R2) could have been used to obtain that resistance. However, should I(V1) have been used, the total resistance would have been negative.

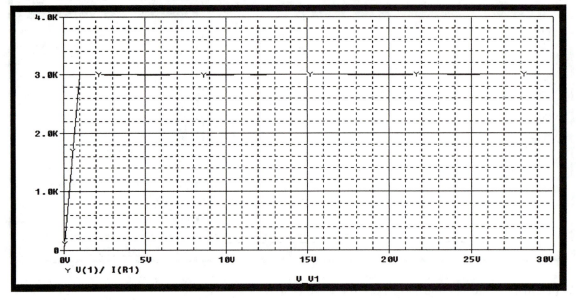

Let us plot the power delivered to **R1** and **R2** and that delivered by the source **V1**.

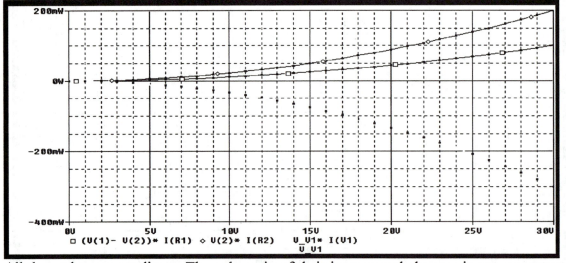

All three plots are nonlinear. Thus, the ratio of their incremental changes is not a constant. For instance, when V(1) increases from 10 volts to 20 volts, the power to **R2** increases by about 60 milliwatts. As V(1) increases from 20 volts to 30 volts, the power to **R2**

increases by about 120 milliwatts. Also, the plot shows that the source **V1** has a negative powerflow. It is delivering power to the resistors. Note however, that for any value of V(1), the sum of the power delivered to **R1** and **R2** is equal to the power delivered by the source **V1**. For instance when the voltage of V1 is 30 volts, the total power delivered to **R1** and **R2** is 300 milliwatts. This is equal to the power delivered by the source **V1**.

THE PRINCIPLE OF SUPERPOSITION

In any linear network containing several independent sources, the voltage across and the current through any resistor may be calculated by finding the algebraic sum of all the voltages and current caused by each independent source in the circuit. All other independent voltage sources are replaced by short-circuits. All independent current sources are replaced by open circuits. Any resistances associated with the sources must be included in the circuit.

What's new?
1. Labeling of axis
2. Enter text on **PROBE** plots

We shall use the circuit in Figure 2.02 to demonstrate the principle of superposition.

Figure 2.02

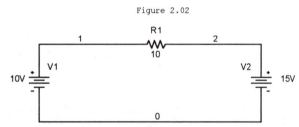

To begin, we perform a 10 second **transient analysis** with both **V1** and **V2** connected into the circuit.

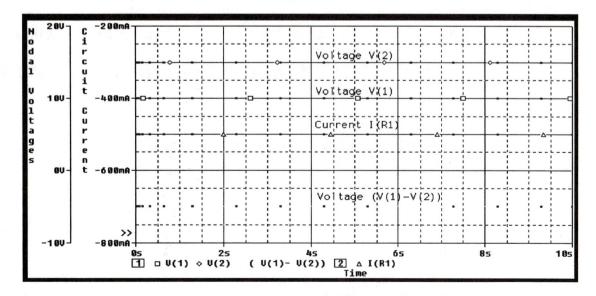

The **PROBE** plot shows the three nodal voltages and the current. The nodal voltages V(1) and V(2) are equal to the sources voltages **V1** and **V2**. Their values are 10 volts and 15 volts respectively. The voltage across **R1, (V(1)-V(2))** is –5 volts because of the voltage difference between V(1) and V(2). The current I(R1**)** has a value of –500 milliamps because it is flowing into the negative terminal of **R1**.

To help in the identification of the different voltage and the current traces, labels were affixed to them.

To do this:
1. In the **PROBE** task bar, click on the **ABC** icon. The **Text Label** box opens.
2. In **Enter text label,** type the desired label and click on **OK**.
3. Drag to the desired location on screen and click to place. The **Text Label** box with desired label typed is shown :

To allow a viewing of the voltages and the current traces on one **PROBE** plot, despite the large difference of their amplitudes, two Y-axes were used. The procedure for obtaining the second Y-axis was covered in the preceding chapter. The labeling of the Y-axis is new.

1. With the **PROBE** screen open, click on **Plot** and click on **Axis Settings.**
2. In **Axis Settings** box, click on **Y-Axis** tab.
3. In the **Y-Axis Number** box keep 1.
4. In the **Axis Title** box, type the desired label. Click on **OK**.
5. Get a second **Y-axis**.
6. Repeat this procedure, in the **Y-Axis Number** box, replace **2** with the desired label.

We shall next run an analysis of Figure 2.02 with **V2** set to zero volts.

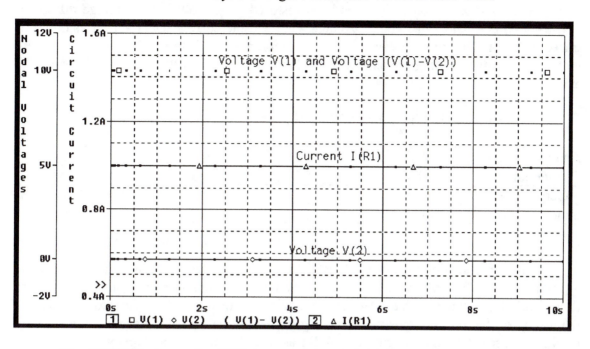

The **PROBE** shows that V(2) is zero volts. V(1) is at 10 volts, as is (V(1) -V(2)), i.e. V(1) minus V(2). The current I(R1) is one amp. It flows in a clockwise direction into the positive terminal of **R1**. We next run an analysis with **V1** set to zero volts. The results are shown below.

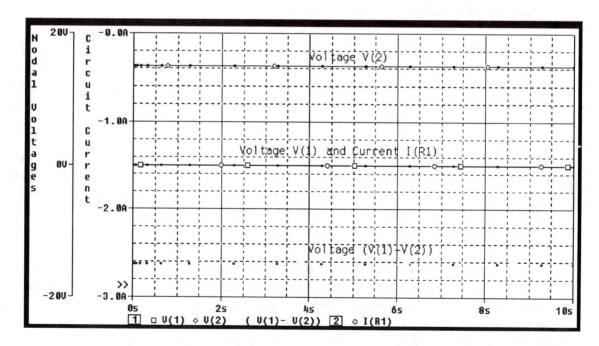

V(1) is at zero volts. V(2) has been restored to 15 volts. The voltage difference (V(1) -V(2)) is equal to –15 volts. Voltage V(1) and the current I(R1) appear as only one trace on the **PROBE** plot; however, they are referenced to different **Y-axes**. V(1) is zero volts on the **Nodal Voltage** Y-axis and I(R1) is –1.5 amps on the **Circuit Current** Y-axis.

Let us summarize the results of the three runs.

Analysis Mode	V(1)	V(2)	(V(1)-V(2))	I(R1)
Conventional	10 V	15 V	-5 V	-.5 A
Superposition	10 V	0 V	10 V	1 A
Superposition	0 V	15 V	-15 V	-1.5 A

The superposition method enables us to obtain the circuit currents and voltages due to each source. The voltage source **V2** causes a counterclockwise current of 1.5 amps to flow, while the voltage source **V1** causes a clockwise current of 1 amp to flow. If we add the results of the two superposition runs algebraically, we obtain the same value for both the voltage across and the current through **R1** as with the conventional analysis.

THEVENIN'S THEOREM

What's new ?
1. performance analysis

In many complex circuits, our interest is often centered on one particular component, the load resistor. It is the payoff in human terms. The load resistor does what we want. It may be an electrical heat appliance, a light bulb, a loudspeaker, or a myriad of other appliances. The rest of the circuit can be thought of as the necessary means to bring about the desired end.

From the perspective of the load resistor, any network connected to it, regardless of the number of independent or dependent current or voltage sources, the number of resistors connected , all this complexity can be reduced to a single voltage source in series with a single resistor. These two elements are defined as the **Thevenin voltage source** and the **Thevenin resistor** and comprise the **Thevenin equivalent circuit**.

The **Thevenin voltage** is that voltage that one would measure at the network nodes of the **load resistor** with that resistor removed. This voltage is referred to as the open circuit voltage. The **Thevenin resistor** is that resistance measured at the same nodes with the load resistance removed and all voltage and current sources, both of the independent and the dependent type in the circuit set to zero volts and zero amps.

The formal statement of **Thevenin's theorem** applied to dc circuits reads:

Any two-terminal, linear bilateral dc network can be replaced by an equivalent circuit consisting of a voltage source and a series resistor.

We shall use Figure 2.03 to demonstrate the theorem.

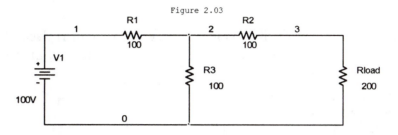

Figure 2.03

Determining the Thevenin voltage

We replace **RLoad** with a 1000 MΩ resistor. This approximates an open circuit condition at node 3. The result of a 10 second **transient analysis** is shown on the **PROBE** plot.

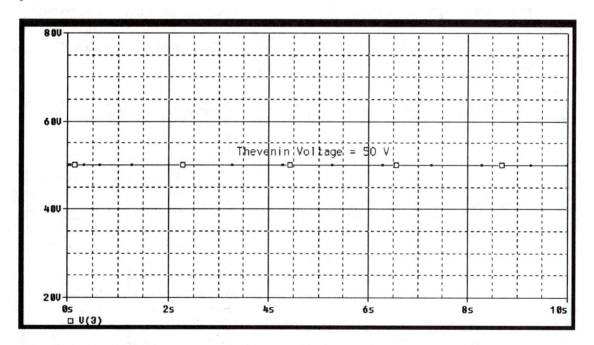

The **Thevenin voltage** is shown to be 50 volts. In the circuit of Figure 2.03, if **RLoad** were removed and a voltmeter were placed between node 3 and node 0, the voltmeter would read 50 volts.

Determining the Thevenin resistor.

We replace **RLoad** with a variable resistor **Rvar**. We measure V(3), the voltage across that resistor, as a function of **Rvar** as its resistance changes. In our analysis, **Rvar** changed from 100 Ω to 200 Ω in steps of 5 Ω. When the voltage across **Rvar** is one-half of the **Thevenin voltage**, the resistance of **Rvar** is equal to the resistance of the **Thevenin resistor** of Figure 2.03.

To obtain a plot of the voltage V(3) across **RLoad** replaced by **Rvar**, we do a **Performance analysis.** It allows us to plot V(3) as a function of the changing values of **Rvar** or **RVAL** as PROBE plots it.

Proceed as follows:
1. With the **Probe** screen open, click on **Plot**. Click on **Axis Settings.**
2. In the **Axis Settings** box, click on **X Axis**.
3. Click on **Performance Analysis** and click on **OK**. The X-axis will be in units of **RVAL**.
4. Click on **Trace** and click on **Add Trace**. The **Add Traces** box opens.
5. In the **Add Traces** box, in **Functions or Macros**, click on **Max(1).** It will appear in the **Trace Expression** box. It will plot V(3) up to its maximum value.
6. In **Simulation Output Variables**, click on V(3). It appears in the **Trace Expression** box.
7. Click on **OK**. The trace of V(3), as a function of **RVAL** is plotted on the **PROBE** screen. **RVAL** is the resistance value of **Rvar**.

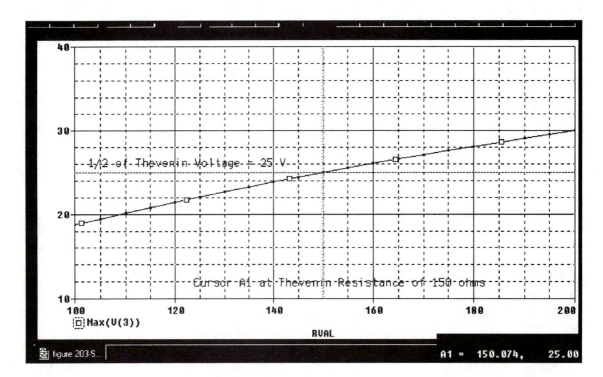

From the data obtained, the **Thevenin equivalent circuit** is shown in Figure 2.04.

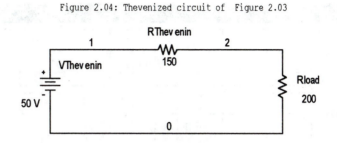

Figure 2.04: Thevenized circuit of Figure 2.03

This circuit will cause the same voltage across and current through **RLoad** as the circuit in Figure 2.03. The reader is encouraged to run a **PSpice** analysis using Figure 2.03 and Figure 2.04 to compare their currents and voltages.

The reader is cautioned to note that the **Thevenin voltage** is not across **RLoad.** It would appear only at node 2 with **RLoad** removed.

While for this simple circuit the application of **Thevenin's theorem** may seem unnecessary, such is not the case when more complex circuit are to be analyzed. The savings in analytic effort can be considerable. To demonstrate the point, let us obtain the **Thevenin equivalent circuit** of Figure 2.05.

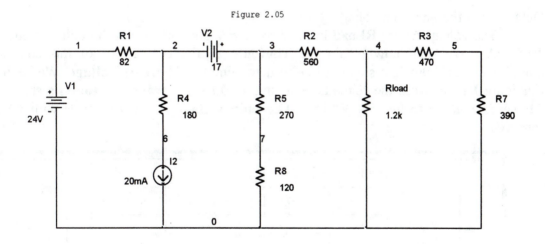

Figure 2.05

We shall first run a **transient analysis** of 10 seconds duration with the circuit as is. This will give us the voltage across and the current through **RLoad.** The result of the analysis is shown on the **PROBE** plot.

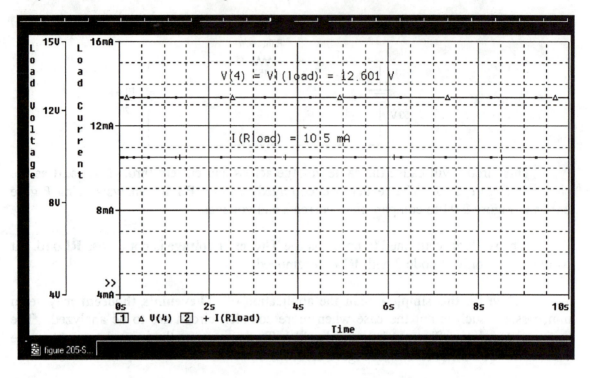

Determining the Thevenin voltage

The voltage across **RLoad** is equal to 12.6 V and the current through it is equal to 10.5 mA. We shall now obtain the **Thevenin equivalent circuit** of this somewhat complex circuit. Our first step shall be to determine the **Thevenin voltage**. We begin by removing **RLoad** from the circuit. Running a 10 second **transient analysis** shows the **Thevenin voltage** to be 16.4 V. The **Thevenin voltage** is equal to V(4) with **RLoad** removed.

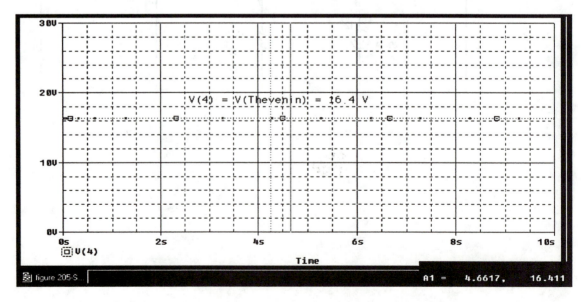

Determining the Thevenin resistance

We remove **RLoad** and place a variable resistor across nodes 4 and 0. We perform a **Parametric** analysis. The range for **Rvar** was from 200 Ω to 500 Ω in increments of 50 Ω. On the **PROBE** plot, the cursor was placed at one half the **Thevenin voltage** at 8.2 volts. The corresponding value of **RVAL**, which is the value of the **Thevenin resistance,** is 395 Ω.

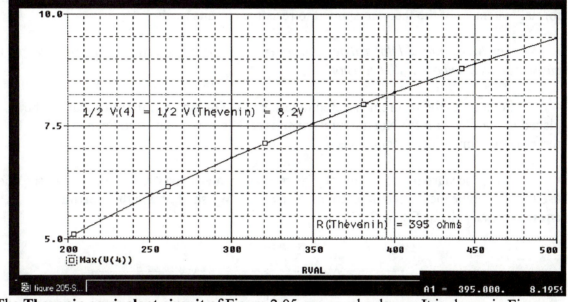

The **Thevenin equivalent circuit** of Figure 2.05 can now be drawn. It is shown in Figure 2.06.

Figure 2.06: Thevenin equivalent circuit of Figure 2.05

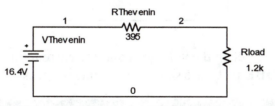

Running a **PSpice** analysis of this circuit produced a voltage across Rload and a current through it that are identical to those obtained for the original circuit of Figure 2.05. Verifying this statement is left as an exercise for the reader.

A comparison of the load voltages and the load currents of Figures 2.05 and 2.06 shows that they are in reasonable agreement. What is important in this analysis is the relative ease by which the **Thevenin equivalent circuit** can be obtained by **PSpice** when compared to methods of traditional analysis. Also, should it be desired to change **RLoad** and to obtain its new voltage and current, the **Thevenin equivalent circuit** quickly leads to the desired answers.

NORTON'S THEOREM

It has been shown that from the perspective of **RLoad,** any network to which it is connected can be replaced by a voltage source in series with a resistor. We shall demonstrate next that from the perspective of **RLoad**, any network can be represented by a current source, called the **Norton current source**, in parallel with a resistor, called the **Norton resistor**. These two elements comprise the **Norton equivalent circuit.** The formal statement of the theorem applied to resistive circuits reads:

Any two-terminal bilateral dc network (resistors are bilateral) can be replaced by a current source and a parallel resistor.

Sometimes the theorem is stated in terms of a **Norton conductance** that can be placed in parallel with the **Norton current source**. We recall that conductance and resistors are dimensional reciprocals.

The procedure for obtaining the **Norton equivalent circuit** is analogous to the one used to determine the **Thevenin equivalent circuit**. We shall obtain the **Norton equivalent circuit** for Figure 2.07.

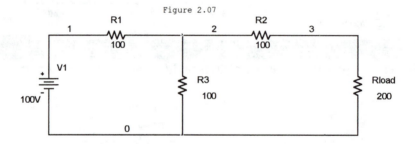

Figure 2.07

First we shall obtain the voltage at and the current through **RLoad** using the above circuit. Their **PROBE** plots are shown. The voltage across **RLoad** is 28.6 volts and its

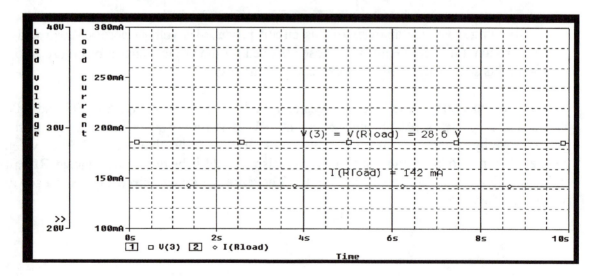

current is 142 milliamps. We shall use that data, as we did in the previous example, to compare the circuit of Figure 2.07 with its Norton equivalent.

Determining the Norton Current source

To obtain that source, we replace **RLoad** in Figure 2.07 with a short circuit. A small resistor, **Rshort**, of 1e-10 Ω will do fine. This format results in a resistor of value $1*10^{-10}$ Ω. The result of a 10 second **transient analysis** is shown on the **PROBE** plot. It shows that the **Norton current** flowing through **Rshort** is equal to 333.3 milliamps. The short-circuit current is an important parameter for utility engineers who must determine this current at the entry point into a customer's premises. The value of that current determines the size of the fuses or circuit breakers needed.

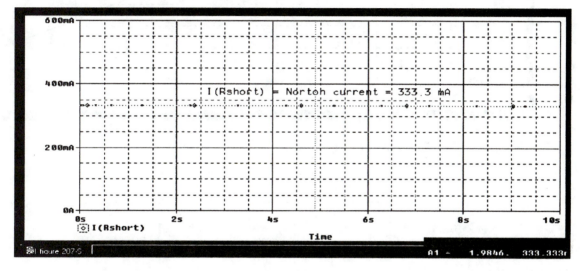

Determining the Norton Resistance

The procedure is similar to the one used to obtain the Thevenin resistance. In Figure 2.07, we replace **Rshort** with **Rvar**. We vary our **Rvar**, or its value **RVAL** as shown on the **PROBE** plot, until the current through it is equal to one-half the **Norton current**. At that current, **Rvar** is equal to the **Norton resistance.** We set the values of **Rvar** to vary from 100 Ω to 200 Ω in steps of 5 Ω. A **parametric** and a **performance analysis** were done. The **PROBE** plots show the result. When I(R4), which is the current through **RVAL,** is equal to 166.7 milliamps,   which is equal half of the **Norton current,** the value of **RVAL** is equal to 150 Ω. The **Norton equivalent circuit** can now be constructed. No doubt , the reader is aware that an increasing **RVAL** causes a decline in the current I(R4); hence the negative slope of the current plot in the **PROBE** plot below.

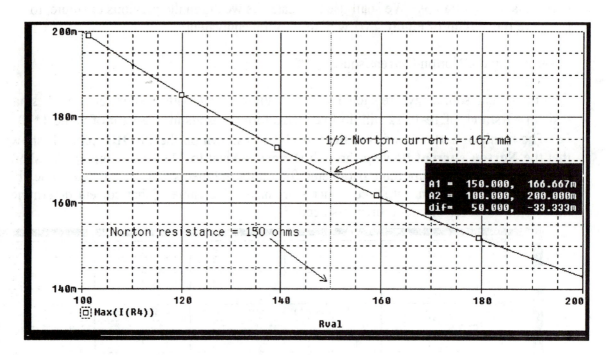

The **Norton equivalent circuit** is as shown:

Figure 2.08: Norton equivalent circuit of Figure 2.07

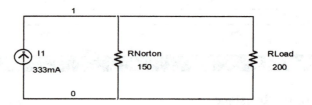

The **Norton current source** has been placed so that the voltage **V(1)** is positive. The reader is encouraged to run **a PSpice** analysis to verify the equivalency between Figure 2.07 and Figure 2.08.

SOURCE CONVERSION

In the derivation of the **Thevenin** and the **Norton equivalent circuits**, we started with the same original circuit. Let us examine the relationship that exists between these two. Examining the data from **Figures 2.03** and **2.08**, we observe that the **Thevenin resistance** and the **Norton resistance** are both equal to 150 Ω. If we divide the **Thevenin voltage** of 50 volts of Figure 2.03 by the **Thevenin resistance** of Figure 2.03, we obtain 333 milliamps, the **Norton current source** in Figure 2.08. Similarly, if we multiply the **Norton current** in Figure 2.08 by the **Norton resistance** in that figure, we obtain the **Thevenin voltage**, 50 volts, of Figure 2.03. Thus, we can convert either circuit into the other. The choice is often one of utility. In circuits consisting primarily of parallel components, the **Norton equivalent circuit** is the one of choice. In circuits consisting of circuit elements mostly connected in series, the **Thevenin equivalent circuit** is the preferred one.

Source conversion is a general concept. Any voltage source with a series resistor can be replaced by a current source with that resistor in parallel. The value of the current source will always be the ratio of the original voltage source divided by the resistance of its series resistor. The reciprocal conversion also holds.

MAXIMUM POWER TRANSFER THEOREM

In an electrical system, the voltage and/or the current sources are the sources of electrical power. The load resistor(s) is (are) the primary recipient of that power. The various elements connected between the electrical source and the load resistor(s) serve primarily as the transmission media.

An Ideal Voltage Source in a Circuit

An ideal current or voltage source is a source that can transmit an infinite amount of power, should such be required. The voltage of an ideal voltage source is independent of the amount of current drawn from it. The current of an ideal current source is independent of the voltage across its terminals. We shall use circuits that contain these ideal sources to demonstrate these characteristics.

Let us investigate the power flow to **RLoad** as it changes from 0 Ω to 500 Ω in steps of 50 Ω of Figure 2.09.

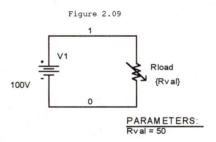

Figure 2.09

We need to perform a **parametric** and a **Performance analysis** as was done previously. The relationship between the power drawn from **V1** and the value of the variable **RLoad** is shown on the **PROBE** plot. It shows that for small values of **Rval**, the power drawn from **V1** tends toward infinity. The circuit is approximating a short circuit condition. Increasing **Rval,** in effect approximating open circuit conditions, reduces the power drawn from **V1** toward zero watts. It is important to realize that during the entire range of **Rval**, the voltage **V1** remains at 100 V.

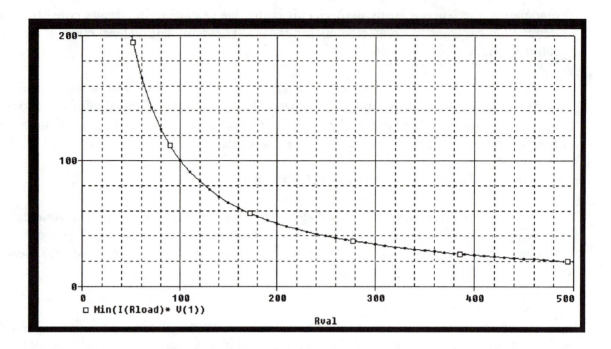

A Nonideal Voltage Source in a Circuit

A real voltage source can be represented by an ideal voltage source in series with a source resistor. The smaller that resistor, the more closely the nonideal voltage source resembles an ideal voltage source. The addition of a source resistor profoundly changes the power flow in a circuit. We modify Figure 2.09 to study the power flow to a resistor **RLoad** from a non-ideal source. The modified circuit, containing **Rsource** of 100 Ω is shown in Figure 2.10.

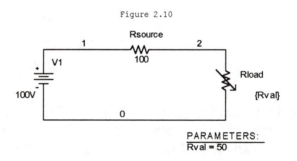

Figure 2.10

Rval is made to vary from 10 Ω to 500 Ω in steps of 10 Ω. A **parametric** and a **performance analysis** yielded the result shown next.

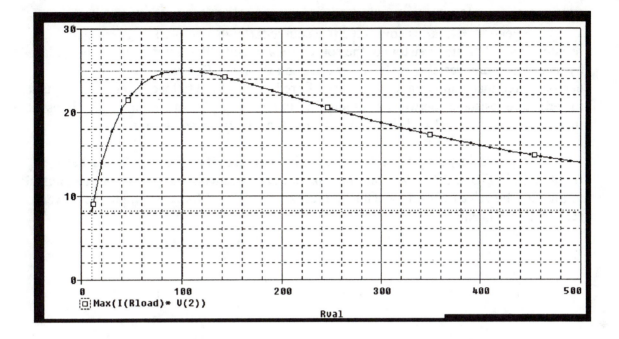

The power to **RLoad** depends upon V(2), I(Rload) and **Rval**. For the extreme values of **Rval**, either its voltage or its current is small, limiting the amount of power received by **RLoad**. However, somewhere between the extreme values of **Rval**, there is a setting for **Rval** at which the power to **RLoad** is at a maximum. That power is 25 watts and it occurs when **Rval** = 100 Ω. That value is equal to **Rsource**! This result is expressed formally in the **maximum power transfer theorem.**

For a load connected directly to a dc voltage supply source, maximum power will be transferred from the source to the load when the load resistance is equal to the internal resistance of the source.

A Fallacy to Guard Against
        A superficial reading of this theorem may lead one to believe that in order to extract maximum power from a source, a source resistor is needed, to which a load resistor is then matched. We recall that in the absence of **Rsource**, the power we could extract from the ideal voltage source that we investigated in the circuit of Figure 2.09 tended toward infinity watts at decreasing values of **Rval**. Even for the range of **Rval** used, that power went off the **PROBE** plot at 200 watts. Presently, the maximum power that we could extract was 25 watts. The proper interpretation of the theorem is that given the ever-present **Rsource**, the most for which we can hope is to extract the maximum possible power by matching **RLoad** with **Rsource**.

Thevenin's and Maximum Power Transfer Theorems Combined

In general, **RLoad** is connected to a source through a network of connected resistors, including any resistors of the sources in a circuit. The latter could all be replaced by their **Thevenin resistor** without any effect on **RLoad.** But we have just demonstrated that maximum power is transferred from source to load if **RLoad** is equal to the effective resistance between itself and the source. Thus, for maximum power transfer from source to load, **RLoad** should be equal to the **Thevenin resistance.**

To prove the point, the **Thevenin equivalent circuit** of Figure 2.03 has been reproduced in Figure 2.11. We shall run a **parametric analysis** and **performance analysis** to find out the value of **Rval**, which is the resistance value of **Rvar**, at which the power to **RLoad** is at a maximum.

Figure 2.11: Thevenin equivalent circuit of Figure 2.03

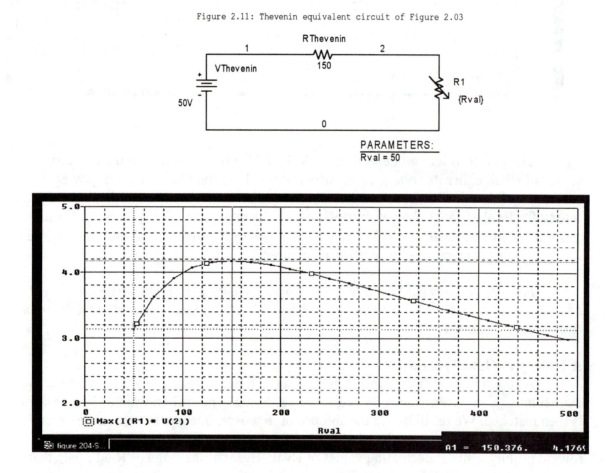

The **PROBE** plot proves that maximum power of 4.18 watts is transferred from source to load when **Rvar** = **Rval** = 150 Ω. It is nice when all works out well!

Problems

2.1   For the circuit shown, verify the superposition theory for resistor **R3.** Proceed by
        getting the voltage and current due to either voltage source. Verify the theorem
        with a conventional analysis with **V1** and **V2** in the circuit.

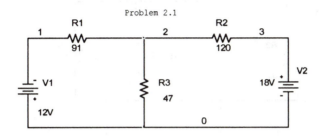

2.2   Use superposition to find the voltage across and the current through each
        resistor. Verify the results by running a **PSpice** analysis with both the voltage
        and the current sources connected in the circuit.

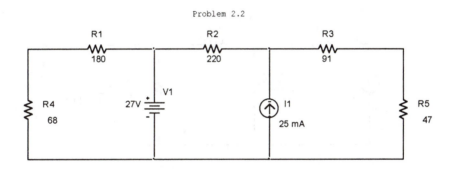

2.3   **V1** can change from 10 V to 40 V in steps of 1 V. Use superposition to find the
        voltage across and the current through resistor **R3**. Use a **PROBE** plot to show the
        voltage across **R3** as a function of **V1**. Verify your results by running a **PSpice**
        analysis with both voltage sources active in the circuit.

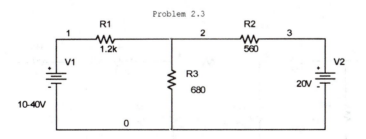

2.4   Use superposition to find the voltage across and the current through resistor **R3**.
        The **E1** source has a forward voltage gain of .5. Verify your results by running a
        PSpice analysis with both independent voltage sources in the circuit. Note: There is

no connection between the crossover point of node 1 and node 0 at the left
terminal of **R1**.

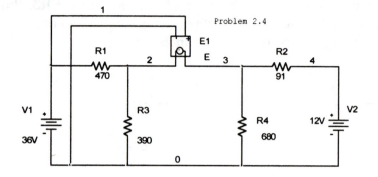

2.5   Use superposition to find the voltage across and the current through resistor **R2**.
       Verify your result by running a **PSpice** analysis with both independent current
       sources in the circuit. The **F1** source has a current gain of 10.

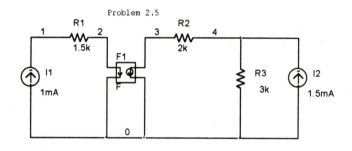

2.6   Find the **Thevenin equivalent circuit** relative to **RLoad**. Verify your
       results by running a **PSpice** analysis of the original, unthevenized circuit.
       Compare the results of the two analyses to insure the equivalency
       of the two circuits.

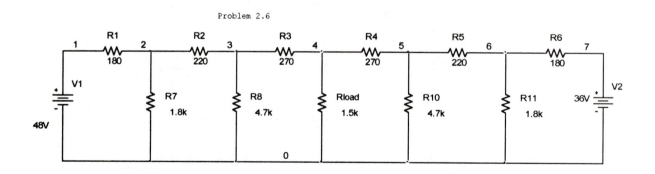

2.7   Find the **Thevenin equivalent circuit** relative to **RLoad**. Check your answer by comparing your design data with that of the original circuit.

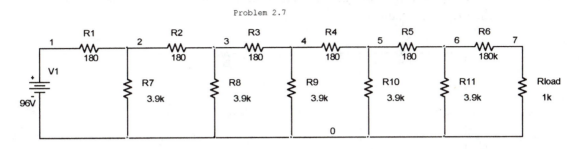

Problem 2.7

2.8   The circuit contains an independent voltage source and a current-controlled current source. Find its **Thevenin equivalent circuit** with respect to **RLoad**. Verify your analysis by comparing the voltage across **RLoad** of the original with that of the **Thevenin equivalent circuit**. The current gain of the **F** source is .5.

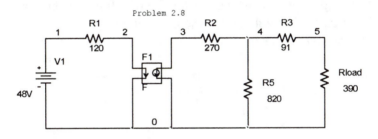

Problem 2.8

2.9   Find the **Norton equivalent circuit** with respect to **RLoad**. Verify your results by comparing the voltage across **RLoad** and its current with that of the original circuit.

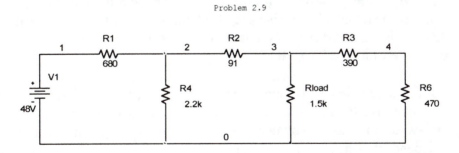

Problem 2.9

2.10   Change the **Norton equivalent circuit** of Problem 2.9 into a **Thevenin equivalent circuit**. Compare the voltage across and the current through **RLoad** of the two circuits to insure their equivalency.

2.11   Find the **Norton equivalent circuit** with respect to **RLoad**. Verify your results
by comparing the voltage across and the current through **RLoad** with that of
the original circuit.

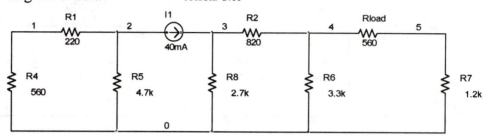

Problem 2.11

2.12   The circuit contains a variable resistor **R1** that can change its value from
$100\ \Omega$ to $2000\ \Omega$ in steps of $50\ \Omega$.
Find the following:
(a) The resistance of **R1** for which its power is at a maximum.
(b) The value of its voltage and current at maximum power
(c) The range of the resistance of **R1** for which its power is at least 6 W.
(d) The range of its voltage for the condition in part (c).
(e) The range of its current for the condition in part (c).

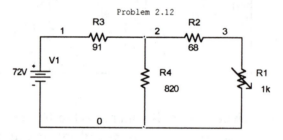

Problem 2.12

2.13   Answer all the questions in Problem 2-12 with the following changes: in
(a), use the range of the resistance of **R1** for which its power is at least 1 watt.
Also, it is left to the reader's initiative to pick the range and the incremental steps
of the resistance value of **R1**.

2.14   For the circuit shown, use superposition to find the nodal voltage and the
current through each resistor. Specify the contribution that each voltage source
makes to the nodal voltages and the currents. Run an   analysis with both
independent voltage sources **V1** and **V2** in the circuit and compare the results of
the two analyses. The voltage gain of the **E1** source is .4.

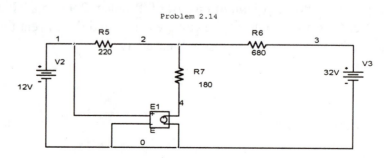

Problem 2.14

2.15   This circuit contains a current-controlled current source **H1**. Use superposition to
         get all nodal voltages and the current through each resistor. Run a conventional
         analysis with both independent sources **I1** and **V1** in the circuit and compare the
         results of the two runs. The current gain of **H1** is 10.

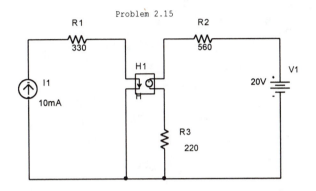

Problem 2.15

TRANSIENTS IN RC AND RL CIRCUITS

TRANSIENT RESPONSE OF AN RC CIRCUIT

What's new?
1. The capacitor and the inductor in PSpice
2. The voltage pulses **VPULSE and VPWL**
3. The **EXP** (exponential) function
4. Initial condition of a capacitor
5. Power and energy of a capacitor
6. **TD** (time delay)

A circuit is in the transient mode when it adjusts from one energy level to a different one. This shift could be from a state of zero energy to a nonzero state, from a nonzero state to a zero energy state or from an initial energy state to a new one. In either case, during the transient phase, the equations that govern circuit behavior are often exponential in kind. At the completion of the transient phase, the circuit variables resemble the applied energy source in their mathematical format.

We shall investigate the transient response of the circuit in Figure 3.01. **V1** is a rectangular voltage pulse **VPULSE**, with its parameters listed and its **PROBE** trace shown.

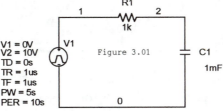

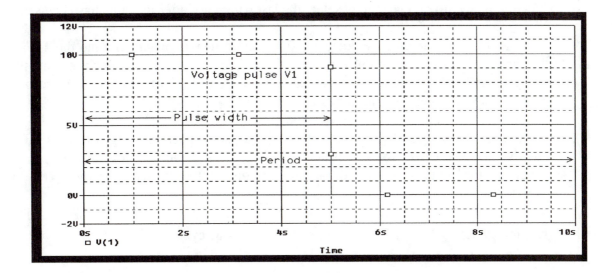

To obtain **VPULSE**, click on **Place part** icon. In **Place Part** box, select Source and scroll to **VPULSE.** Click on it to select it. Click on OK. Place VPULSE and click to deselect, ending placement. **V1** and **V2** are the voltage levels of the pulse. **TD** is the time delay from zero time**. TR** and **TF** are the rise and fall times of the pulse. They need not be equal to each other. **PW** is the pulse width and **PER** is the period of the pulse. Setting these parameter values as shown results in the following voltage pulse. To set the value of these parameters, proceed as follows:

1. On the **Capture screen**, double click on a parameter to select it. The **Display Properties** box opens.

```
┌─────────────────────────────────────────────────────────────────┐
│ Display Properties                                          [×]  │
├─────────────────────────────────────────────────────────────────┤
│                                        ┌─Font──────────────────┐ │
│   Name:  V1                            │ Arial 7 (default)     │ │
│                                        │                       │ │
│   Value: [0V          ]                │ [ Change... ] [ Use Default ] │
│                                        └───────────────────────┘ │
│   ┌─Display Format──────────┐   ┌─Color──────────────────────┐   │
│   │  ○ Do Not Display       │   │ [ Default            ][▼]  │   │
│   │  ○ Value Only           │   └────────────────────────────┘   │
│   │  ● Name and Value       │   ┌─Rotation───────────────────┐   │
│   │  ○ Name Only            │   │  ● 0°        ○ 180°        │   │
│   │  ○ Both if Value Exists │   │  ○ 90°       ○ 270°        │   │
│   └─────────────────────────┘   └────────────────────────────┘   │
│             [   OK   ]     [  Cancel  ]     [  Help  ]            │
└─────────────────────────────────────────────────────────────────┘
```

2. In the **Value** box, type the desired value for selected parameter.
3. Click on **OK**. The value appears on the **Capture screen**.
4. Select the next parameter and repeat this procedure until all parameters are set.
5. Click on the **Capture screen** to deselect the last parameter after its value has been set.

The resulting shape of the pulse is a function of the parameters. Should the **TR** and **TF** be increased, a trapezoidal pulse will result. It is even possible to obtain a triangular pulse by the selected setting of the pulse parameters. A **Transient analysis** of 10 seconds duration displays one period of all the circuit variables.

During the 5 second positive portion of **V1**, the capacitor charges toward 10 V. During the next 5 seconds, when **V1** is at 0 V, the capacitor discharges toward 0 V. The mathematical expressions for these two portions of the capacitor voltage V(2) are given by:

Charging cycle: $V(2) = V1(1 - e^{-t/RC})$ volts   Discharge cycle: $V(2) = V1e^{-t/RC}$ volts

The traces of the voltage source **V1**, of the resistor voltage V(1,2) and the capacitor voltage V(2) are shown next.

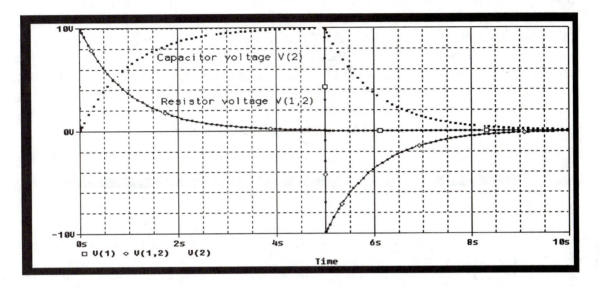

The product **RC** is defined as the time constant of the circuit. It governs the rate at which the capacitor either charges or discharges. For the circuit in Figure 3.01, the product **RC**, often symbolized by the Greek letter τ (tau), computes to 1 second. This means that in 1 second, the capacitor charges to 63% of its final value of 10 volts, in our case 6.3 volts. It discharges in 1 second to 37% of its initial value of 10 volts, or 3.7 volts. This can be verified by direct reference to the above **PROBE** plot. It takes 5 time constants for the capacitor voltage to go from its initial to its final voltage.

As the capacitor voltage increases, the resistor voltage decreases at the same rate. As the capacitor voltage decreases, the resistor voltage increases at the same rate. At any instant of time, the sum of V(1,2) and V(2) is the voltage of **V1**.   The voltages display symmetry relative to the **Y**-axis.

There is one important difference between the two component voltages. The capacitor stores electrical energy. The amount of energy it stores is proportional to the square of the voltage across its plates. To change even a minute amount of energy in zero time is physically impossible since this would require an infinite power flow. No real physical device can do this. In consequence, the voltage across a capacitor cannot change discontinuously. Thus, at t = 0 s and again at t = 5 s, whereas the voltage pulse **V1** can change discontinuously, V(2) does not. This is a fundamental constraint in a circuit that has capacitors. The resistor does not store either electrical or magnetic energy but dissipates energy. Thus, the voltage across it and the current through it can change discontinuously. This is shown by the **PROBE** plot above.

The **PROBE** plot below shows the relationship between the power flow and the energy storage of the capacitor while the source voltage **V1** is applied. An integro-differential relationship exists, which makes the ordinate of the power plot proportional to the slope of the energy plot. An initial inrush of power causes the largest rate of change in the energy content of the capacitor. As the capacitor reaches its fully charged state, its power flow declines toward zero watts. As **V1** reverses polarity at 5 seconds, there is a large power flow outflow accompanied by a decline in the energy content of the capacitor. As the power flow approaches zero watts, the energy of the capacitor approaches zero joules. The capacitor has been drained of all energy. Consequently, its voltage V(2) is zero volts. For any subsequent pulses of **V1**, events will repeat themselves.

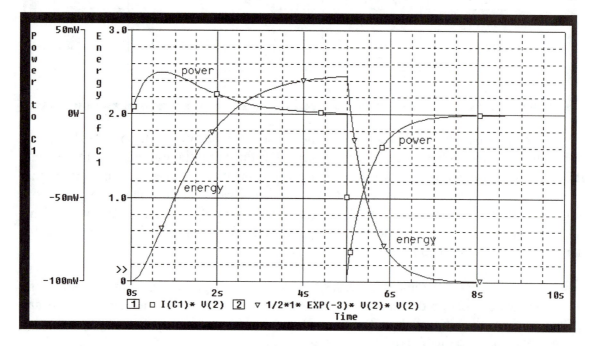

The mathematical expression for the energy content of the capacitor is:

$$W = \tfrac{1}{2}\, C\, V(2)^2 \quad \text{joules}$$

In the **PROBE** plot, the capacitance of 1 millifarad appears as 1***EXP**(-3). The **EXP** (exponential) expression comes from the **Functions or Macros** scroll box of the **Add Traces** dialog box. The square of the voltage V(2) is expressed as V(2)* V(2).

The relative amplitudes of the time constant τ and the pulse width of **V1** determine the shape of V(2).   The pulse width was equal to 5 time constants. This allows the capacitor to reach 10 volts during charging and zero volts during the discharge cycle.   In the **PROBE** plot below, τ was changed to .5 seconds by reducing the capacitance of **C1** to .5 millifarad. The resultant **PROBE** plot is shown.

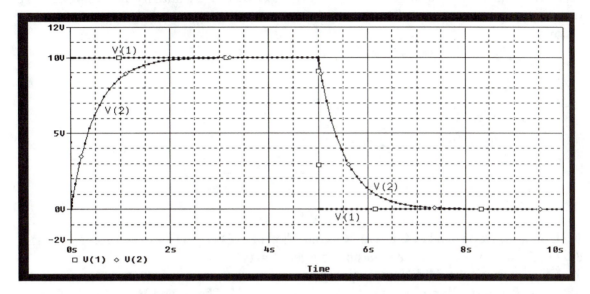

It shows that V(2) reaches its final stage much more quickly. As a result, V(2) more nearly resembles the nodal voltage V(1), which is equal to the source voltage **V1**. We shall next increase τ to 2 seconds by doubling the capacitance of **C1**.   Again, the resulting **PROBE** plot is shown.

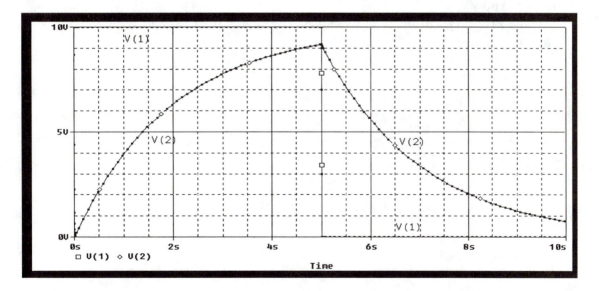

V(2) does not reach 10 volts nor does it completely go to zero volts during one cycle of **V1**.   By changing the value of the time constant τ, we can significantly alter the wave form of V(2). This has important applications. Had the resistance of **R1** been divided by 2 or multiplied by 2, the results on the wave forms of V(2) would have been the same.

In Figure 3.02, the capacitor has an initial voltage of zero volts. How is the capacitor voltage V(2) changed in response to **V1** when V(2) initially is five volts? To obtain that result, we set 5 volts as an initial condition on the capacitor. Proceed as follows:

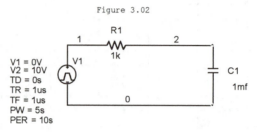

Figure 3.02

1. Double click on symbol **C1**. **The OrCad Capture –[Property Editor]** screen opens.
2. On the bottom of the screen, be sure to select **Parts**.
3. Click on **New**; the **Add New Property** box opens.
4. In the **Property Name** box, type **IC** and click on **OK**. The legend **IC** will appear on the sheet.
5. Below the legend **IC**, type **5V.**
6. Click on **Display**. The **Display Properties** box opens.
7. Select **Name and Value**. Click on **OK**.
8. Click on **Apply**.
9. Close the **Property Editor**. If all went well, the legend **IC = 5V** will appear on the **Schematic** screen.

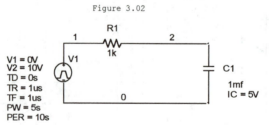

Figure 3.02

Specify a **transient analysis** of 10 seconds duration and run **PSpice.**

The **PROBE** plot shows the results of the analysis.

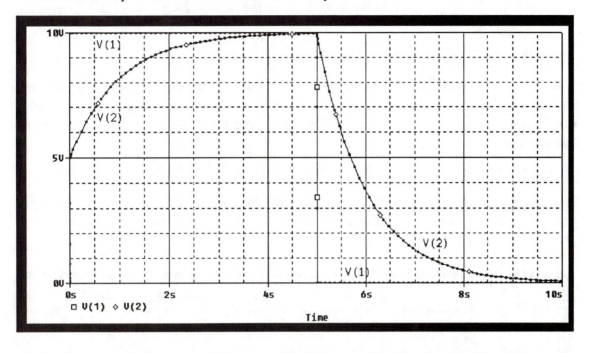

The capacitor voltage V(2) starts rising from its initial voltage of 5 volts toward 10 volts as before. Regardless of the initial capacitor voltage, it still takes 5 time constants to reach 10 volts This is a point well worth remembering! During the second half of **V1**, V(2) is identical to that of the previous run. For any subsequent pulses of **V1**, V(2) will be identical to that of the earlier run.

The RC Circuit as an Integrator

The configuration of the **RC** circuit we have studied is often referred to as an integrator when the voltage across its capacitor is considered the output voltage.
We shall use Figure 3.03 to study its property.

Figure 3.03

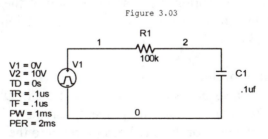

The component values were chosen to reflect those often found in electronic circuits. The value of the time constant τ relative to the pulse width and period of **V1** is important for the proper operation of this circuit.

The current I(R1) or I(C1) is proportional to the rate of change of V(2).

Expressed mathematically: $I(C1) = C\dfrac{dV(2)}{dt}$ amps

By keeping that rate of change small, we assure ourselves of a relatively constant current. This requires that we make our time constant τ much larger than the pulse width of the source voltage **V1**. Given our values of **R1** and **C1** in Figure 3.03, τ computes to 10 milliseconds.

Solving the above equation for V(2) we obtain: $V(2) = \dfrac{1}{C}\int I(C1)dt$ volts

Since I(C1) is relatively constant by design, we can expect V(2) to be a ramp voltage. We perform a **transient analysis of** 2 milliseconds duration and obtain the **PROBE** plot shown.

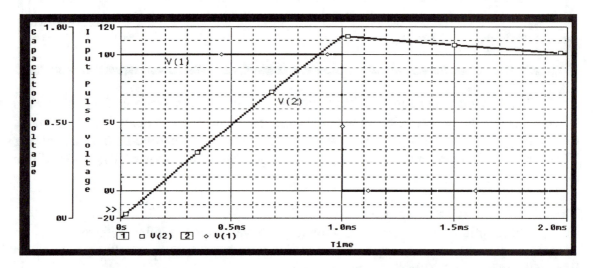

Our prediction was true: during the positive portion of the 10 volt step voltage of **V1**, V(2) is a ramp function with a positive slope. A ramp function is the integral of a step function. Remember Calculus 101!

The RC Circuit as a Differentiator

The **RC** circuit can also be used to differentiate an input voltage. Figure 3.04 shows the traditional configuration of a differentiator circuit. The differentiated input voltage will appear across the resistor that has been placed into the output voltage position.

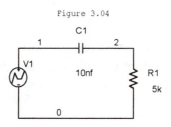

Figure 3.04

For this circuit: $V(2) = RC \dfrac{dV(1,2)}{dt}$ volts

By making the product **RC** small compared to the period of the input pulse **V1**, we assure that V(1,2) approximates **V1** in a short time, hence their derivatives become identical and the voltage V(2) is proportional to the derivative of the input voltage pulse **V1**. For our circuit, the product **RC** is equal to 50 μs.

The symbol for **V1** is new. It is found in the **SOURCE** library as **VPWL**. This is a piecewise-linear voltage pulse. We shall use it to generate a triangular wave form for **V1**. It will rise from zero volts to ten volts in 1 millisecond. It will decline toward zero volts in .01 milliseconds. No zero second fall-time is allowed. Such is not realizable in either **PSpice** or in nature.

There are no pulse parameter entrees found on the **Schematic** screen. They must be set as follows:

1. Double click on the symbol for **V1**. The **OrCad Capture-[Property Editor]** dialog box opens.
2. Selecting **Parts** tap, on the bottom of the **Property Editor**, scroll to right until the setting boxes for **T** (time) and **V** (voltage) appear.
3. Enter all parameter values as shown below. For **T3** enter 1.01 milliseconds; for **T5** enter 2.01 milliseconds.
4. Click on **Apply.**
5. Click to close the **Property Editor**.

V1 has a period of 1 millisecond. This is far longer that the 50 microseconds of the time constant τ for the circuit of Figure 3.04. A **transient analysis** of 2 milliseconds will yield two cycles of **V1,** as shown on the **PROBE** plot.

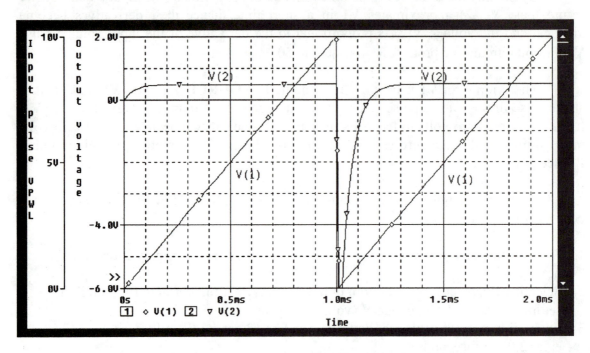

V(2) is a step function . This is the derivative of the ramp function of the voltage pulse **V1**. It is only at the transition times of **V1** that V(2) departs from being a step function. This is due to the short, yet finite, time in which the capacitor charges and discharges. Choosing an even smaller time constant would reduce these times and V(2) would more closely resemble a step function. Our objective has been attained.

Pulse Train Applied to a RC Circuit

To the circuit in Figure 3.05, we shall apply **V1** having the parameters shown. To see ten cycles of that pulse, we need to run a **transient analysis** of 4 milliseconds duration.

Figure 3.05

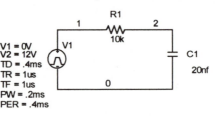

The results of the analysis are shown in the **PROBE** plot below. For the first time we have introduced a time delay **(TD)** into our analysis. On the **PROBE** plot, **V1** and V(2) do not begin till .4 milliseconds have elapsed. V(2) migrates upward. The question is, where to? The answer is until the average capacitor voltage V(2) is equal to the average voltage of **V1**. That voltage is equal to 6 volts. **V(2)** has achieved 6 volts   by about 1.5 ms.

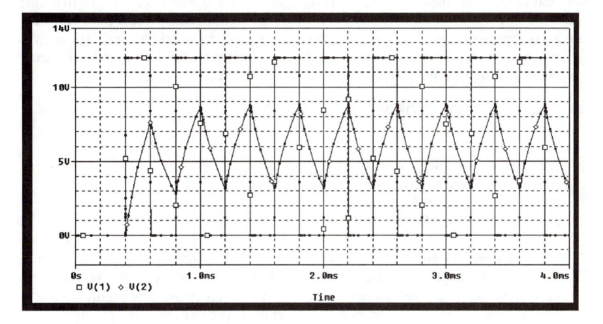

For reasons of clarity, the resistor voltage V(1,2) has been plotted below on a separate **PROBE** plot, as shown.

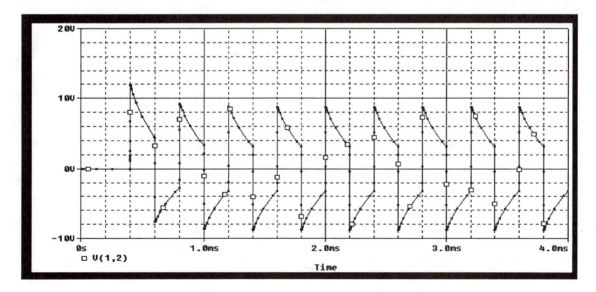

While the capacitor voltage migrated upward, the resistor voltage migrates downward. Where to? It will migrate until its average voltage is zero volts! Of special interest is the length of the migration period. In both cases V(1,2) and V(2) begin their journeys at t = .4 milliseconds. By the time t = 1.5 milliseconds, both voltages have attained their constant values of 6 volts for V(2) and zero volts for **V(1,2)**. But why did it take about 1 millisecond for the two voltages to stabilize themselves? The answer rests with the value of the time constant. Its value is equal to 200 microseconds. For five time constants, that amounts to 1 millisecond. Adding .4 milliseconds of **TD** to that time brings us approximately to 1.5 milliseconds. The time during which the variables migrate is defined as the *transient response* of the circuit. The time after which the migration is completed is defined as the *steady-state response* of the circuit.

TRANSIENT RESPONSE OF AN RL CIRCUIT

What's new?
1. The inductor in **PSpice**
2.   Initial condition of the inductor
3. Power and energy of the inductor
4. The use of multiple **PROBE** plots

The transient response of the **RL** circuit shown in the circuit of Figure 3.06 is our next objective. We apply the **VPULSE** of Figure 3.01 to a resistor **R1** with a resistance of 10 Ω and an inductor **L1** with an inductance of 10 H. In this circuit, any energy stored will be magnetic energy. It will reside in the inductor. This energy manifests itself as an initial current flowing through **L1**. For now, we stipulate that there is no initial energy contained in the circuit. Thus, we set the initial condition on the inductor equal to 0 A. The procedure is identical to that for the capacitor.

Figure 3.06

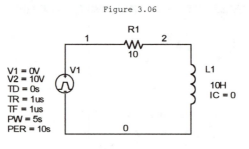

We perform a **transient analysis** for 10 seconds. The results are depicted on the **PROBE** plot.

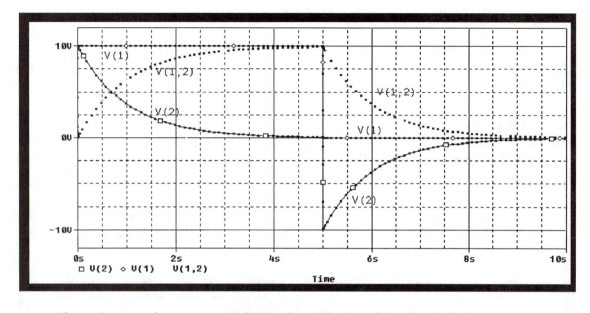

If we compare the present **PROBE** plot with that of the circuit if Figure 3.01, we find that the plots of V(1,2) and V(2) have been interchanged. Since I(L1) is proportional to the energy content of the inductor **L1**, the current I(L1) cannot change discontinuously. Since V(1,2) is proportional to I(L1), it cannot change in zero time either.

The rate of charging and discharging for the present circuit is determined by its time constant τ. It is calculated as the ratio of **L1/R1** seconds. Given the values of **R1** and **L1**, this ratio computes to 1 second. This was also the value of τ in Figure 3.01. The equations governing the charging and the discharging portions of the inductor are of the same format as those for the capacitor.

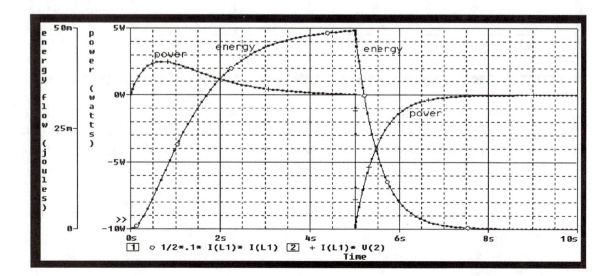

The expression for the energy content of an inductor is:

$$W = \tfrac{1}{2}\, L\, I(L1)^2 \quad \textbf{joules}$$

The power and the energy for one cycle of **V1** are shown on the preceding page. The plots are of the same shape as those for Figure 3.01. In this case, the inductor current I(L1) plays the same role as the capacitor voltage V(2) in Figure 3.01.

The preceding **PROBE** traces show that a positive power flow increases the energy content of the inductor. Conversely, as the power flow becomes negative, the energy content of the inductor declines. These curves are analogous to the **RC** circuit of Figure 3.01.

We shall next investigate the relationship between V(2), the relative amplitudes of the time constant and the pulse width of **V1**. To start, we change the time constant to .5 seconds by increasing **R1** to 20 Ω. The result on V(2) is shown below.

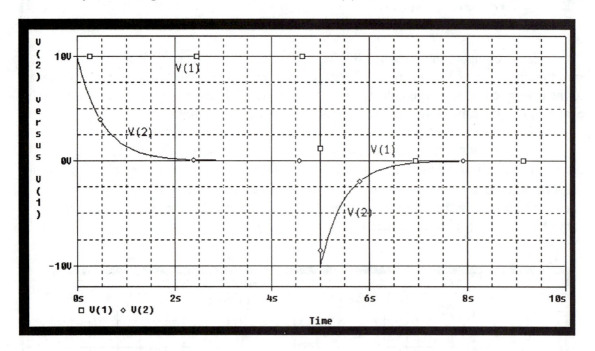

With the time constant equal to .5 seconds, it takes about 2.5 seconds for V(2) to reach zero volts, compared to 5 seconds when the time constant was equal to 1 second . If the time constant is increased to 2 seconds by setting **R1** equal to 5 Ω, V(2) will be changed as shown.

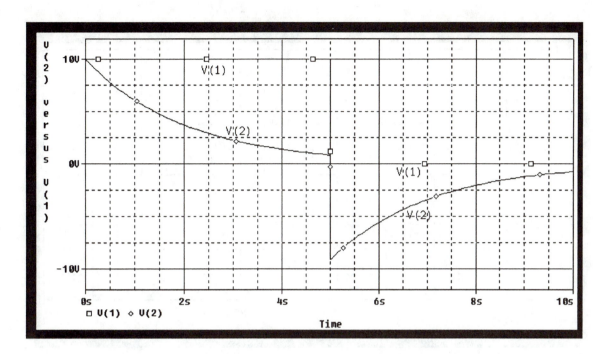

The voltage V(2) does not go completely to zero volts. To do so, with a time constant of 2 seconds, would take 10 seconds. However, **V1** changes every 5 seconds. If we compare the effect of changing the time constant on V(2) for this circuit and for the circuit in Figure 3.02, we note that the results on V(2) are analogous. In each case, a smaller time constant makes for a faster transition time of V(2) . An increase of the time constant makes for a slower transition of V(2).

An initial current of .5 amps flows though the inductor in the circuit of Figure 3.07. This initial condition is set in the same way as for the capacitor in Figure 3.02. The only difference now is that we click on the symbol of the inductor. At the completion of the process, the initial condition should appear on the **Schematic screen** as shown for Figure 3.07. It does!

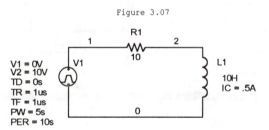

Figure 3.07

A **transient analysis** of 10 seconds yielded the **PROBE** traces shown. For reasons of clarity, the trace inductor current I(L1) is shown first, separate from the resistor voltage V(1,2) and the capacitor voltage V(2).

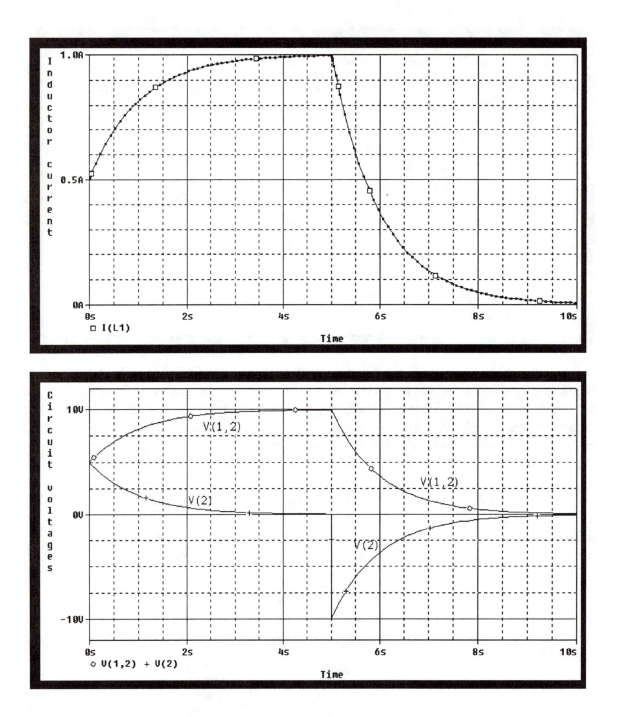

The exponential rate of change for all variables is the same as those of Figure 3.06 since all of them are governed by the same τ of 1 second. The current I(L1) begins its migration from an initial current of .5 amps toward its terminal value of 1 amp. The resistor voltage V(1,2) and the capacitor voltage V(2) both start their journeys at 5 volts. This is a requirement of Kirchhoff's voltage law: the sum of these two voltages must be equal to the source voltage **V1** at any time. They do! It bears repeating: the exponential rates are not affected by the initial condition of the inductor's current.

The RL Circuit as an Integrator

The circuit in Figure 3.08 is used to demonstrate the RL circuit as an integrator circuit. **V(2)** is the output voltage of the circuit.

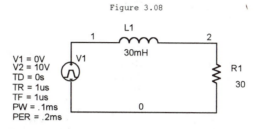

Figure 3.08

The following equations govern the relationship between the circuit current I(L1) or I(R1), the inductor voltage V(1,2) and the resistor voltage V(2). The time constant **L1/R1** for the circuit is 1 millisecond. The pulse width of **V1** is .1 millisecond.

$$V(1,2) = L\frac{di}{dt} \text{, therefore } i(t) = \frac{1}{L}\int V(1,2)dt, \text{ and } V(2)= i(t)(R1)= V(2) = \frac{R}{L}\int V(1,2)dt$$

If we make the time constant of the circuit much larger than the pulse width of **V1**, V(1,2) will change slowly over time. This keeps V(1,2) close to **V1**. Thus, in the above integral, we can replace V(1,2) by **V1**. Therefore, V(2) is proportional to the integral of **V1**. This we shall demonstrate. We shall run a **transient analysis** of 400 us duration. In practice, the author has found that 100 data points give a reasonably clear plot. Thus, a time step of 400 μs/100 points = 4 μs print step was used. The results of the data run are shown next.

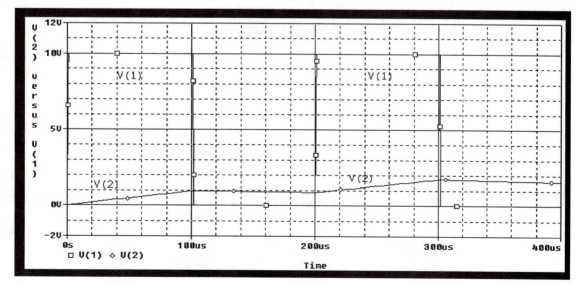

The traces show that during the positive step of 10 volts of **V1**, the voltage V(2) is a positive ramp function. Such a function is the integral of a step function.

The RL Circuit as a Differentiator

The circuit in Figure 3.09 is used to demonstrate the **RL** circuit as a differentiator. The time constant of the circuit is equal to 130 nanoseconds.

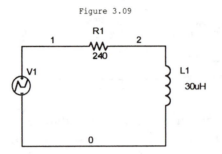

Figure 3.09

The input voltage pulse **V1** is a triangular pulse. Its parameters are set as shown in the **Property Editor** dialog box below.

Source Package	T1	T10	T2	T3	T4	T5	T6	T7	T8	T9	V1	V10	V2	V3	V4	V5	V6	V7	V8	V9	Value
VPWL	0		1m	2m	3m	4m					0		4	0	4	0					VPWL

By proper design of this circuit, the voltage V(2) will be the derivative of the input voltage **V1.** V(2) will be proportional to the time derivative of V(1,2) if the time constant is small compared to the pulse width of **V1.** The small time constant of 130 nanoseconds assures that. We have:

$$V(2) = L\frac{di}{dt} \quad \text{and} \quad i(t) = \frac{V(1,2)}{R1}$$

For small τ, **V(2)** reaches its extreme but small values. This makes **V(1,2)** approximately equal to V(1). Therefore, we substitute:

$$i(t) = \frac{V(1)}{R1} \text{ amps } \text{ which yields: } V(2) = \frac{L1}{R1}\frac{dV(1)}{dt} \text{ volts}$$

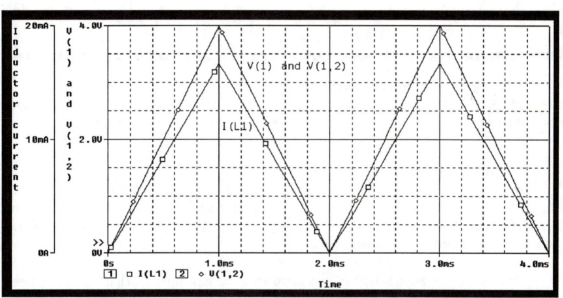

Thus, V(2) is proportional to **V(1)**! All of this is shown on the **PROBE** traces above.

The **PROBE** plot above made use of multiple plots. To obtain these:

1. Click on **Plot**. A drop-down menu opens.
2. In it, click on **Add Plot to Window**. The additional plot opens. The symbol SEL>> appears, indicating that it is the active plot.
3. Enter plots on it in the usual manner.

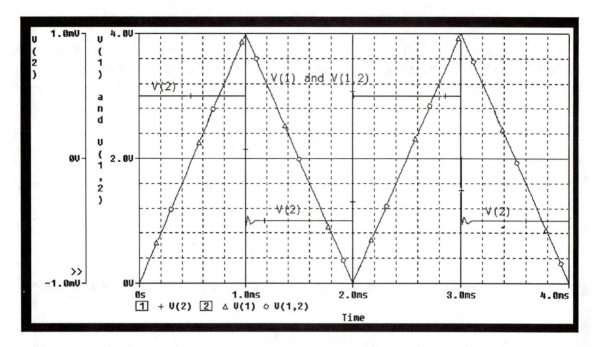

The amplitude of V(2) is much less than that of **V1**. However, the objective is to show that V(2) can be made the derivative of **V1**. In practice, often wave shaping is more critical than amplitude. It is relatively easy to amplify a voltage to a desired level.

Current Pulse Applied to an RL Circuit

In the work of the electrical engineer, current or voltage pulses are often applied to circuits that are more complex than the pulses we have studied so far. By means of the piecewise-linear **(PWL)** current or voltage pulses, we can approximate complex wave forms and study a circuit's response to them.

The circuit in Figure 3.10 is energized by a current pulse. That pulse has been modeled by a **PWL** current pulse. We shall show graphically the integro-differential relationship that exists between the inductor's current and voltage.

Figure 3.10

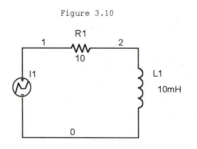

The plot of **I1** is shown on the **PROBE** plot as **I(I1)**. Its data pairs of time and current must be entered into the **Property Editor** sheet as was done for Figure 3.09.

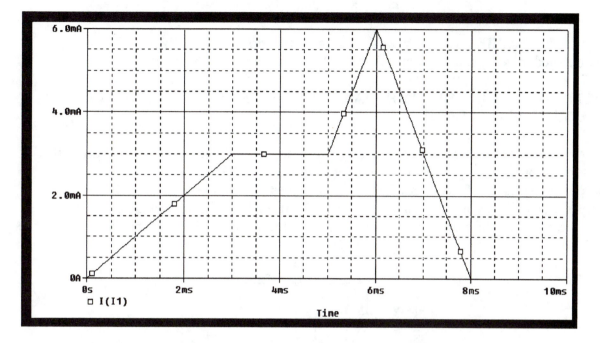

The **PROBE** plots of the inductor's current and voltage are shown.

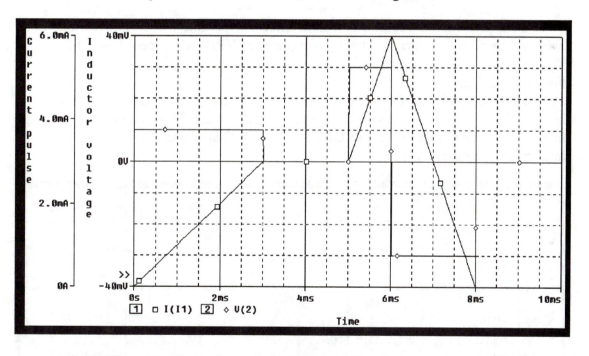

The inductor's voltage is proportional to the derivative of its current. Let us analyze the plots for $0 < t < 3$ ms.

$$V(2) = L\frac{di}{dt} = 10mH\frac{(3mA - 0mA)}{(3ms - 0ms)} = 10mV$$

The **PROBE** trace for V(2) shows its voltage to be ten millivolts. The reader is encouraged to verify the inductor voltage V(2) for any other time interval.

DETERMINATION OF THE AVERAGE VALUES OF CURRENT AND VOLTAGE WAVEFORMS

Often it is desired to obtain average currents and voltages from input sources that have little or no average values. These average values are defined as the dc values of a current or voltage. Circuits are explicitly designed to obtain output variables with dc values. Such circuits are defined as rectifiers. **PSpice** allows for an easy determination of the average value of a current or voltage. Let us recall Figure 3.01, relabeled as Figure 3.11.

Figure 3.11

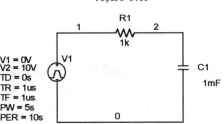

We shall perform a 10 second **transient analysis**. To obtain the average value of V(1), which is equal to the average value of the voltage of **VPULSE**, proceed as follows:
1. Click on **Trace.**
2. Click on **Add Trace.**
3. In the **Add Traces** dialog box, move cursor to the **Functions or Macros** scroll box.
4. Click on **AVG( )** to select it. **AVG( )** will appear in the **Trace Expression** box.
5. Click on V(1). It will be added to **AVG( )** in the **Trace Expression** box.
6. Click on **OK** to generate plot.

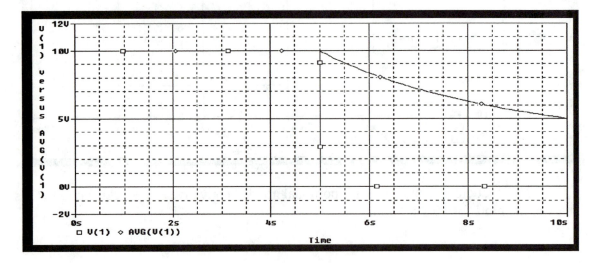

The plot of **V(1)** and its average, **AVG(V1)),** is shown above.

For *t* less than or equal to 5 seconds, the average voltage of V(1) and its instantaneous voltage are the same. However, for *t* > 5 seconds, the voltage of V(1) approaches five volts. This is the average voltage of V(1) over 10 seconds. Its average value could have been easily determined for this simple waveform by inspection, but not every wave form is that simple! The plot of AVG(V(1)) is not a plot of a circuit variable. It is a plot of a mathematical property of the voltage of V(1).

What is the average value of V(2)? Its plot is shown next.

For 0 < *t* < 5 seconds, V(2) increases to 10 volts and AVG(V(2) increases to 8 volts. For *t* > 5 seconds, V(2) declines toward zero volts and AVG(V(2)) declines toward the average voltage of **V1** over 10 seconds. That voltage is 5 volts. We recall that when a pulse train was applied to Figure 3.05, the capacitor voltage V(2) migrated toward the average value of **VPULSE**, 6 volts in that case. Here we see the same migration. The reader is encouraged to determine AVG(V(1,2)).

The circuit in Figure 3.12 has the voltage pulse **V1** applied to it. We perform a **transient analysis** of 100 milliseconds. For that period, **V1** is in effect a pure dc voltage. Our objective is to obtain V(2) and AVG(V(2)).

Figure 3.12

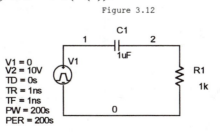

The results of our analysis are shown below. V(1) and AVG(V(1)) have the same value of 10 volts since that is the characteristic of a dc voltage. By contrast, V(2) and AVG(V(2)) are both equal to zero volts after an initial transient period. If we refer back to Figure 3.04, we recall that it was a differentiator circuit as is the present one. Speaking like a mathematician, the derivative of a constant is zero, and so the **PROBE** data shows. Speaking like an engineer, a dc current or voltage is blocked by a capacitor, and so it is!

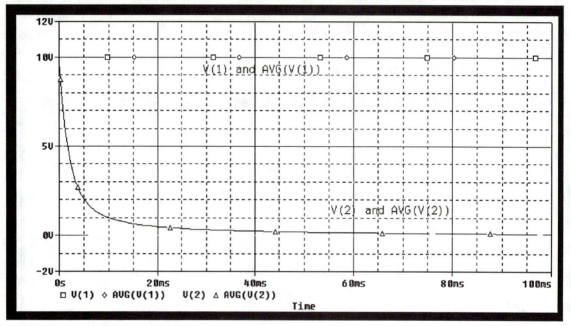

THE SWITCH IN PSPICE

What's new?

1. The use of the **S** switch in **PSpice**

So far in our analysis, we have used pulses to energize our circuits. Now we shall use a switch to open and close a circuit. Study Figure 3.13. Switch **S1** has two sets of terminals. On its left side are the controlling terminals. If the voltage across them is equal to 1 volt, as determined by **V1**, the controlled terminals on the right side are shortened by the closed switch. A near-zero resistance condition exists across them. If the voltage across the controlling terminals is 0 volts, again as determined by **V1**, the switch opens and the controlled terminals have a very large resistance across them, simulating an open-circuit condition .

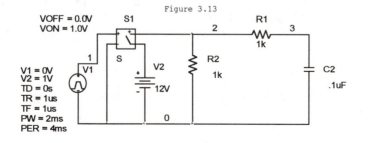

Figure 3.13

Specific to our circuit, if **V1** is 1 volt, the dc source **V2** is connected to the circuit. The capacitor **C2** will charge to 12 volts at a rate determined by the time constant **R1*C2**. If **V1** is zero volts, **V2** is disconnected from the circuit. The capacitor will discharge at a rate determined by the time constant **(R1+R2)*C2**. The large resistance of the open switch **S1** is in parallel with **R2** and has little if any effect on the time constant.

To obtain the switch, proceed as follows:
1. Click on the **Part** icon.
2. Click on **ANALOG.**
3. Scroll to and select **S.**
4. Click on **OK.**
5. Position the switch on Schematic screen. Click to place.
6. Right click to **End Mode**
7. Click on the switch to deselect.

For this circuit, we shall run a 4 millisecond **transient analysis** to obtain the capacitor voltage. The pulse parameters of **V1** are set in the usual manner. The resulting **PROBE** plot is shown next.

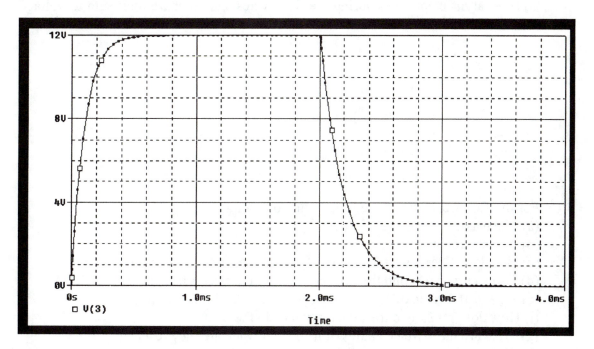

As predicted, the capacitor charged to 12 volts in .5 milliseconds. For the charging cycle, 5τ = 5*R1*C2 = 5*1k*.1uF = .5 milliseconds. For **V1** equal to zero volts, the capacitor discharges toward zero volts in 1 millisecond. For the discharge cycle, 5τ = **5*(R1 + R2)*C2** . This computes to **5*(1k + 1k)*.1uF**= 1 millisecond. Switches can be used singly or in combination in a variety of circuits to achieve desired performance.

Problems

3.1   A pulse train, its parameters as defined below, is applied to the **RC** circuit shown. Print and plot the voltage V(2) across the capacitor, the voltage V(1,2) across the resistor and the voltage V(1) on one plot. Explain the results.

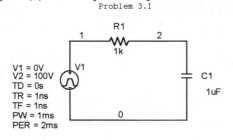

Problem 3.1

3.2   The RC circuit has a voltage pulse applied having the listed parameters:
For this circuit, perform a **PSpice** analysis that will show five cycles of the capacitor voltage V(2) and the and resistor voltage V(1,2). Find the sum of these voltages at different times during the five cycles and compare their sum to voltage **V1.**

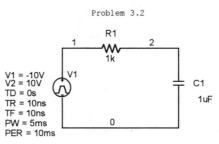

Problem 3.2

3.3   For the **RC** circuit shown, determine the following:
(a) The maximum voltage of V(2).
(b) The maximum voltage of V(1,2).
(c) How long does it take for either of them to reach their maximum voltage?
(d) Using cursor **A1**, find the time constant of the circuit from the plot of V(2).
(e) What is the value of **R** in the expression for the time constant?
(f) How does it relate to the total resistance in the circuit?
(g) What are the extreme values of I(C1) and when are they reached?
(h) Plot the power and the energy contained for the capacitor.
(i) What are their extreme values and when are they reached?

3.4   For the circuit shown, a triangular pulse **V1** is applied with its parameters shown on its **PROBE** plot.

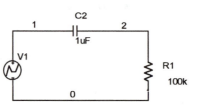

Problem 3.4

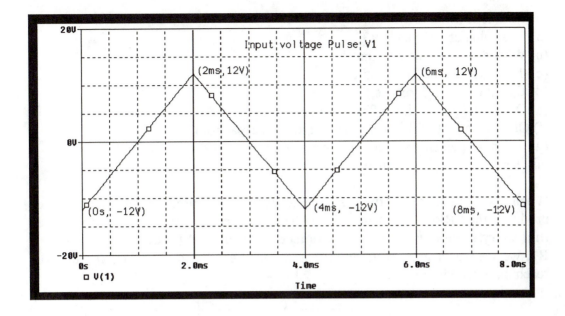

(a) Determine the voltage V(2) and AVG(V(2))
(b) Compare AVG(V(2)) to AVG(V(1))
(c) Explain the action of the capacitor in the circuit
(d) What practical use could this circuit have?

3.5   For this circuit, determine the value of the total resistance in the circuit from the value of the time constant. Remember that **V(2)** goes toward 63 % of its final value in one time constant. Incidentally, is this not a neat way of determining the total resistance of a circuit, no matter how complex?

Problem 3.5

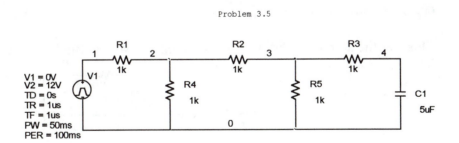

3.6   The circuit consists of two **RC** sections.

Problem 3.6

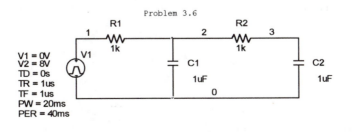

Find the following:
(a) The voltages V(2) and V(3).
(b) What are their time constants?
(c) Find the current I(C1) and I(C2).
(d) Explain their differences.
(e) Obtain the power flow of both capacitors.
(f) Obtain the plot of the energy stored in both capacitors.
(g) What is the energy for both capacitors at t = 20 milliseconds and t = 40 milliseconds? Explain your findings.

3.7   When it was attempted to run a **PSpice** analysis of the circuit shown, the program produced the ERROR message shown. This was because **in PSpice it is required that every node must have a dc connection to ground (node 0).** If we look at **R1**, we see that neither of its nodes has it because **C1** and **C2** block any dc from reaching the nodes of **R1**.

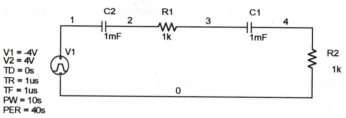

Problem 3.7

ERROR- -Node 2 is floating
ERROR- -Node 3 is floating.

To remedy this situation, place a large resistor in parallel with either **C1** or **C2**. This will provide a dc path for both node 2 and node 3. A sufficiently large resistor will not substantially alter the operation of this circuit. Now that the problem is fixed, find the time constant of this circuit, the voltage V(4) and the current through the circuit.

This was the author's solution to the problem. It worked fine!

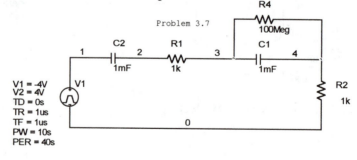

Problem 3.7

3.8   For the integrator circuit shown, find
        (a) The voltage V(2)
        (b) The slope of V(2) for 0 < *t* < 20 milliseconds
        (c) The slope of **V(2)** for 20 milliseconds < *t* < 40 milliseconds
        (d) Calculate the value of V(2) at *t* = 20 milliseconds and compare it with its value
            from its **PROBE** plot.

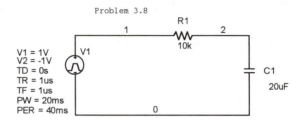

3.9   Some parameters of **V1** from the previous problem have been changed. Answer all
        questions as before. In addition, what, if any, is the difference in the current voltage
        V(2) compared to the previous one?

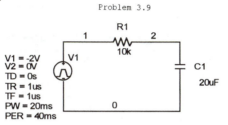

3.10   For this circuit, perform a **transient analysis** of 40 milliseconds duration.
        (a) Find the plot of all the circuit voltages.
        (b) What are the extreme values of these voltages?
        (c) Which of them has the fasted rate of change and why?
        (d) Which one has the slowest rate of change and why?
        (e) Get the power flow for each capacitor.
        (f) Which of them has the largest, and which one has the smallest power flow?
        (g) Get the energy content for each capacitor.
        (h) Which of them stores the most and which one the least energy during the time
            of the **transient analysis**?

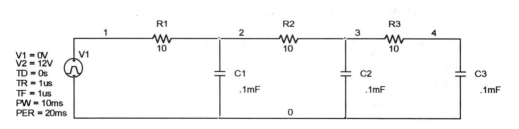

3.11   For the circuit and its input voltage pulse **V1**, perform a **transient analysis** of 200
        milliseconds.
        (a) Find All nodal voltages.
        (b) Explain why some of them change discontinuously at t = 20 milliseconds and
            others do not.
        (c) Obtain all circuit currents.
        (d) Determine the time constant of the circuit from a **PROBE** plot of V(2,3).
        (e) Using the time constant, determine the effective resistance of the circuit.
        (f) Find the power flow of the circuit for all circuit elements.
        (g) Find the energy stored in the circuit.

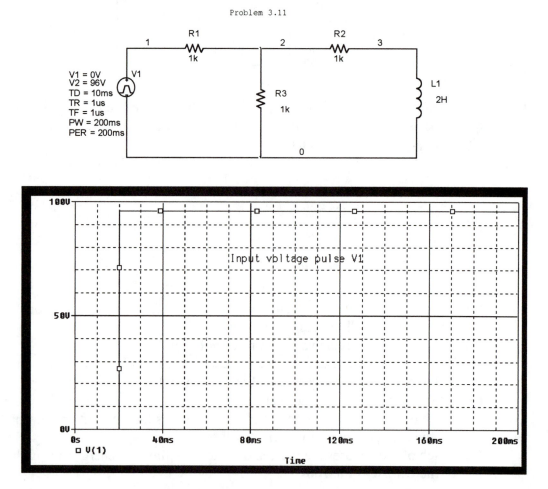

3.12   Use the circuit and its applied voltage pulses **V1** and **V2**, with no initial energy
        stored in the circuit.
        (a) Find the voltage V(2,5) across the inductor.
        (b) Determine when voltage discontinuities occur and their voltage values.
        (c) Find the current through the inductor.
        (d) What is its value at t = 20 milliseconds and t = 40 milliseconds?

(e) Find the power flow to the inductor.
(f) Find the energy stored in the inductor.
(g) Find the two time constants of the circuit.

Problem 3.12

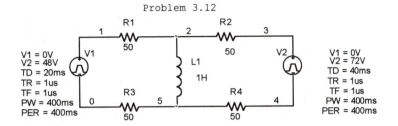

In this problem, apply the **Mirror** command to **V2** so that its parameters appear to the right of its symbol. To do this, you may recall, select **V2**, click on **Edit**, click on **Mirror**, click on **Horizontally**. The parameters will appear as shown.

3.13   For the circuit in Problem 3.12, reverse the polarity of **V2** and repeat the analysis.

3.14   For this circuit, with an initial inductor current of 50 milliamps, find the voltage V(2), the current I(L1), the power flow and the energy stored for the period of the input voltage pulse **V1**.

Problem 3.14

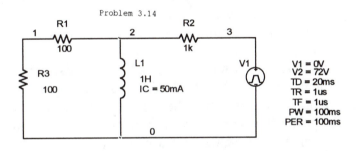

3.15   The circuit has the voltage pulse **V1** shown applied to it. Perform a **transient analysis** of 100 milliseconds duration to obtain all circuit currents and voltages, the power flow and the energy stored in the inductor during 100 milliseconds. Comment on the shape of the circuit variables.

Problem 3.15

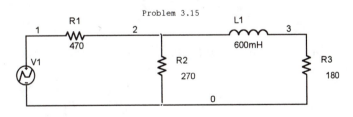

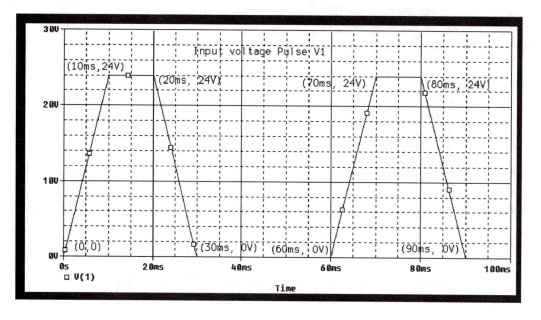

3.16   For the circuit shown below, perform a 2 millisecond **transient analysis.**
        (a) Find the maximum values of V(1,2) and V(2).
        (b) What is the value of the discontinuity of V(2) at $t = 1$ millisecond and why?
        (c) What is the sum of V(1,2) and V(2) for any $0 < t < 1$ millisecond?
        (d) Find the average voltages of V(1,2).
        (e) What is its value at $t = 1$ millisecond and $t = 2$ milliseconds?
        (f) Find the average value of V(2).
        (g) What is its value at $t = 1$ millisecond and $t = 2$ millisecond?
        (h) What is the value of I(L1) at $t = 1$ millisecond and $t = 2$ milliseconds?
        (i) Is the current continuous during the time interval studied and why?
        (j) Obtain the power flow and the energy content of the inductor.
        (k) What are their extreme values and at what time?

Problem 3.16

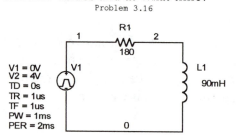

3.17   Repeat Problem 3.16 with the same circuit but with the parameters of **V1** changed
        as shown.   Problem 3.17

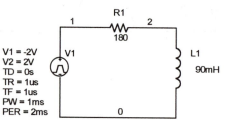

3.18   Repeat Problem 3.16 with the same circuit but with the parameters of **V1** changed
        as shown.

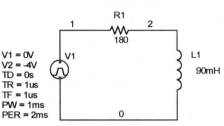

Problem 3.18

In addition, answer the following questions:
(a) Did the transient period for these three problems differ from each other?
(b) Did the wave shapes of the voltages in the three problems change?
(c) How were the average voltages affected by the different input voltage pulses?
(d) Compare the power flows and the energy content for the three problems.

3.19   For the circuit and its input voltage pulse **V1**:
        (a) Find the voltages V(1) and V(1,2) on one **PROBE** plot.
        (b) Put voltage V(2) on a second **PROBE** plot.
        (c) What is the relationship between voltage V(1) and V(2)?
        (d) What is the relationship between V(1) and V(1,2) and why?
        (e) Plot the current I(L1).
        (f) What is its shape compared to V(1,2)? Explain.
        (g) Plot the power flow to **L1.** What are its extreme values? When do they occur?
        (h) Plot the energy content of **L1.**
        (i) What are its maximum and minimum values? When do they occur?
        (j) Are there any discontinuities in the energy plot? If not, why not?
        (k) How would you classify this circuit?

Problem 3.19

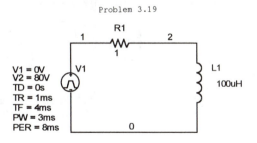

3.20   The circuit has the input voltage pulse **V1** shown applied to it. Find the following:
        (a) The voltages V(1,2) and V(2).
        (b) AVG(V(1,2)) and AVG(V(2)).
        (c) The current I(L1) and AVG(I(L1)).

(d) For reason of scale, use two plots to get the plots of the power flow and
    energy for the inductor.
(e) What are the extreme values of that power?
(f) Explain the shape of the power flow plot.
(g) From the two plots, what is the relationship between the power and the energy
    of the inductor? In particular, explain the relationship of these plots at the time
    of the extreme values of **V1**.
(h) How would you classify this circuit?

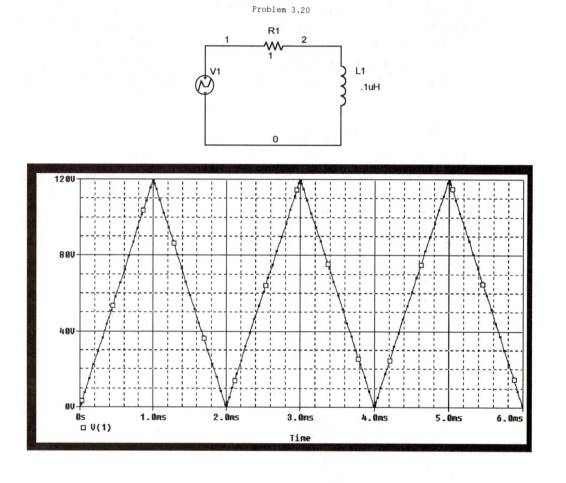

Problem 3.20

3.21 The circuit shown has a current input pulse **I1** applied to it. The inductor carries an
      initial current of 50 milliamps. Perform a **transient analysis** of 6 milliseconds to
      obtain the requested information.

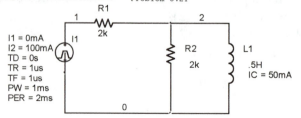

Problem 3.21

For this circuit do the following:
    (a) Plot the voltages V(1,2) and V(2) on one **PROBE** plot.
    (b) Explain their plots.
    (c) On a second **PROBE** plot, plot the sum of V(1,2) and V(2). Compare it to
        voltage **V(1)**. Is Kirchhoff's voltage law satisfied?
    (d) On one **PROBE** plot, plot the currents I(I1) and I(L1).
    (e) What is their relationship?
    (f) On one **PROBE** plot, plot the currents I(R1) and I(R2).
    (g) What is their relationship?
    (h) Compare I(R1) and I(R2) to **I**(I1) and I(L1).
    (i) Plot I(L1) and AVG(I(L1)) on one plot.
    (j) Compare the instantaneous value of I(L1) with its average value.
    (k) Obtain a plot of the power flow of the inductor.
    (l) What are the extreme values of that power flow ?
    (m) When and why do they occur?
    (n) Obtain the energy of the inductor.
    (o) Over the period investigated, does it gain or lose energy?

3.22   This circuit uses two switches with their associated control pulses **V1** and **V3** to
        charge and discharge the capacitor **C1**. The capacitor has no initial voltage across
        it. Perform a **transient analysis** of 400 milliseconds and obtain the voltage V(3)
        across the capacitor. In particular, determine the different time constants of the
        charging and the discharge portion of V(3).

Problem 3.22

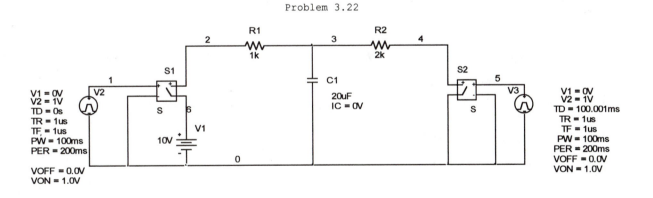

SINUSOIDAL WAVEFORMS
IN RESISTIVE CIRCUITS

APPLYING A SINSOIDAL VOLTAGE SOURCE TO A RESISTIVE CIRCUIT

What's new?

1. The use of the **VSIN** voltage source

We shall next extend our work to include the application of sinusoidal waves to resistive networks. In this chapter, these networks consist of various combinations of resistors only. This was done for ease of analysis. It will allow us to study primarily the behavior and characteristics of sine waves without adding undue analytical complexity introduced by the networks.

Of all the time-varying waves, none are as important and prolific as the sinusoidal current and voltage waves. Most electrical energy is generated by circular motion of conductors in a magnetic field. This results in sinusoidal currents and voltages. The term **ac** usually refers to sinusoidal currents and voltages. Such is not entirely correct. Any waveform that varies with time, such as square waves and triangular waves, is also an **ac** waveform. It is the predominance of the sinusoidal wave that has led to the equivocation of **ac** with sinusoidal waves.

We begin our analysis with Figure 4.01.

Figure 4.01

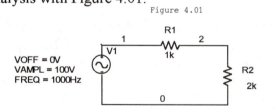

The new component is the voltage source **V1**. To get it:
1. Click on the **Place Part** icon. The **Place Part** dialog box will appear.
2. In the **Libraries** box, scroll to and select **SOURCE**.
3. In the **Part** box, scroll to and select **VSIN** by clicking on it. Click on **OK**.
4. Drag **VSIN** to desired location. Click to place it.
5. Click the right mouse button. Select **End Mode** and click on it.
6. Deselect **VSIN** by placing the pointer anywhere on the **Schematic page** and click.

Three parameters are associated with **V1**. They are set in the usual fashion. The first, **VOFF,** is the value of any offset that **V1** may have. A **VOFF** of zero volts means that **V1** has no dc value and its positive and negative extremes are numerically equal. The second parameter, **VAMPL**, is the absolute peak amplitude of the sine wave. A sine wave has both positive and negative peak amplitudes. The third parameter, **FREQ,** is the frequency of the sine wave measured in Hertz. It is equal to the number of complete cycles per second of the sine wave. Our sine wave has a frequency of 1000 Hz. It goes through a complete cycle 1000 times per second. Conversely, a cycle lasts for 1/ **FREQ** seconds. In our case, that ratio is equal to 1 millisecond. We shall run a **transient analysis** of 2 milliseconds. This will allow the viewing of two cycles of **V1** on its **PROBE** plot. The **PROBE** plot resulting from the analysis is shown next.

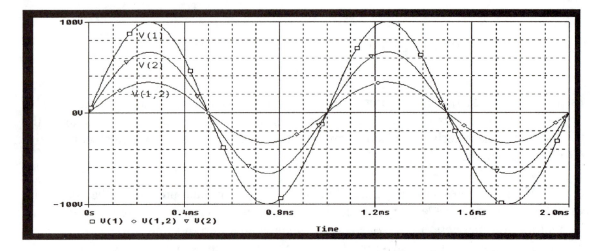

All three voltages reach their positive and negative peaks at the same time. The amplitude of **V1** of 100 volts is set by its parameter. The amplitude of V(1,2) is determined by the expression **R1(R1 + R2)*V1**= 1k/(1k + 2k)*100 = 33.3 volts. The amplitude of V(2) is determined by the expression **R2(R1 + R2)*V1** = 2k(1k + 2k)*100 = 66.7 volts.

All of them have their zero values at the same time. Such sine waves are said to be in phase. Each of their cycles lasts for 1 millisecond. This confirms their frequency of 1000 Hz. The sum of V(1,2) + V(2) is **V1.** Thus**,** the sum of two sine waves having the same frequency is a sine wave. . This is shown on the following **PROBE** plot. The plot of V(1) is identical to the plot of V(1,2) +V(2)**.** This is true for any number of them.

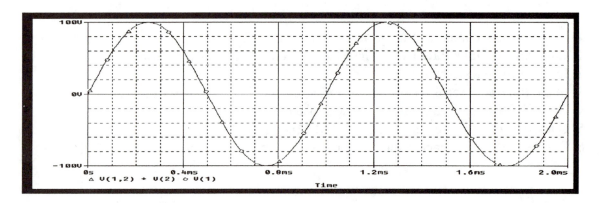

A Problem

All the preceding sine waves cycled through all their different values in 1 millisecond. An oscilloscope can display all their values. No measuring instrument, having a scale/pointer combination or digital readout can measure the instantaneous values of these voltages or any associated current. Even if such values could be measured, the human eye is not quick enough to read these values.

An Attempted Solution

We must find a way of averaging the instantaneous voltages over time. Let us view the average of the above three sine waves.

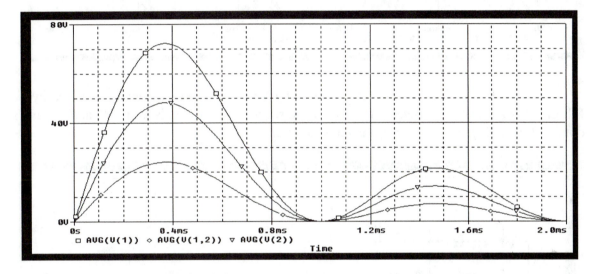

From that plot we observe that for $0 < t < 1$ millisecond, the sine waves do have an average value. We recall from the definition of an average: it is equal to the area under a curve divided by its base. However, for 1 ms $< t < 2$ ms, their average values are declining because more and more negative area is being added. By the time $t = 2$ milliseconds, the net area is equal to zero and so is the average value of our sine waves. Yet, applying **V1** to our circuit, the resistors **R1** and **R2** will heat up. This proves that over time, power is delivered to them. But how can we get power from a voltage that has zero average volts?

A Solution: RMS

The power to a resistor is equal to the square of its voltage divided by its resistance. Let us plot the power to **R1** and **R2**. The following three traces show the instantaneous power delivered by **V1** and the instantaneous power received by **R1** and **R2**. The traces all have a frequency twice that of **V1.** None dips below zero watts. Thus we can expect them all to have nonzero average values.

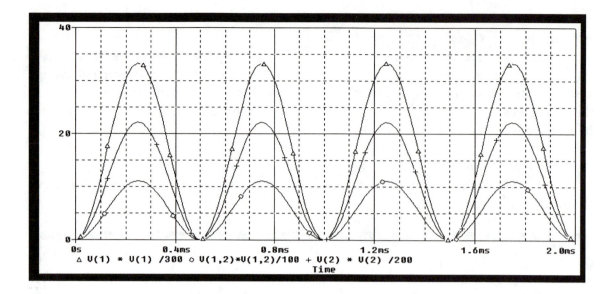

Let us find the average power delivered by **V1** and the average power received by **R1** and **R2**. The average power delivered by the source **V1** and received by **R1** and **R2** as obtained from their **PROBE** traces is tabulated below:

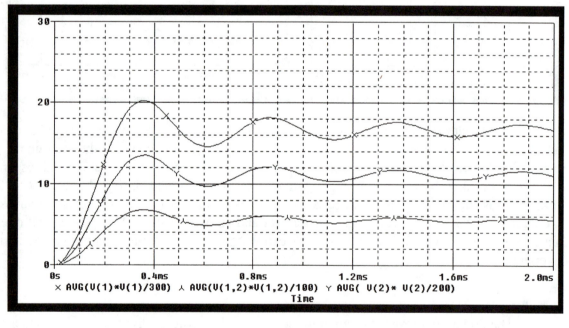

Element	Power (watts)
Source V1	16.67
R1	5.57
R2	11.06

The average power from voltage source **V1** is equal to the average power received by **R1** and **R2**. This is also true for the instantaneous power. Our task is now to find a voltage across **R1** and **R2**, which when squared and divided by the resistance of **R1** and **R2** will yield the time-average power delivered to each resistor. Such a voltage is defined as the **effective voltage**.

$$\text{For } \mathbf{R1}: \quad \frac{V(1,2)effective^2}{R1} = 5.5watts \,.$$

$$\text{Solving for } \mathbf{V(1,2)}_{\text{effective}} = \sqrt{(5.5*100)} = 23.5V$$

The effective voltage across **R1** is 23.5 volts. It will produce the same heating effect in resistor **R1** as would a dc voltage of the same value or a sinusoidal voltage with a peak value of 33.3 volts. If we form the ratio of the effective over the peak voltage across **R1** we obtain:

$$V_{\text{effective}}/V_{\text{peak}} = 23.5 \text{ V}/33.3 \text{ V} = .707$$

This is a general result: for any sinusoidal current or voltage, its effective current or voltage is always .707 times its peak value. The reader is cautioned that this is true only for sine waves. Many a prospective engineering career has been ruined by ignorance of this fact. The reader is encouraged to verify that the effective voltage across **R2** is 47.2 volts.

The **effective voltage** is also defined as the **RMS** value of a sinusoidal voltage. The acronym derives from the procedure by which it is calculated. The original sine wave is squared, then its average value is calculated and finally the square root of that average is obtained to yield the **RMS** value of our original sine wave. For the user of **PSpice**, this process is simplified in that all that needs to be done is to select **RMS** in the **Functions or Macros** scroll box. The result of this procedure is shown next on the **PROBE** plot which shows the three **RMS** voltages of our circuit. At $t = 2$ milliseconds, they have attained their steady values of 16.67 volts, 5.57 volts and 11.06 volts.

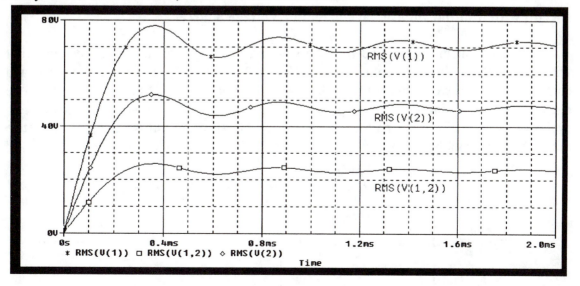

We shall next verify the claim that the power delivered by the **RMS** value of a sine wave to a resistor is equal to the power delivered to the same resistor by a dc voltage source that has the same voltage as the **RMS** voltage of the sine wave. We replace voltage source **V1** with the dc voltage source **V2** shown in Figure 4.02. Its voltage is 70.7 volts, which is the **RMS** voltage of **V1**.

Figure 4.02

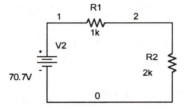

A **transient analysis** was run for 2 milliseconds and produced this **PROBE** plot .

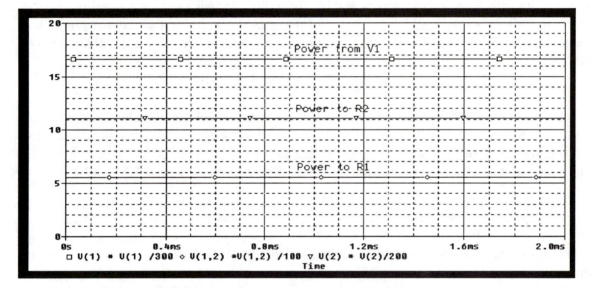

The power values from **V2** and to resistors **R1,** and **R2,** are 16.67 watts, 5.57 watts and 11.06 watts. These are equal to their previously found values.

Circuit with Multiple Sinusoidal Current or Voltage Sources

The circuit in Figure 4.03 has two sinusoidal voltages of the same frequency but with different amplitudes applied to it. We shall investigate the nodal voltages and currents and the power delivered to each resistor. The frequency of 60 Hz means that one cycle of the sinusoidal wave is completed in 1/60 = 16.67 milliseconds. Thus, to see two cycles of any waveform, we perform a **transient analysis** of 34 milliseconds. It is best to mirror **V2** horizontally.

Figure 4.03

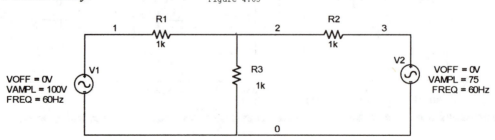

The **PROBE** plots of **V1** and **V2** are shown next. **V1** is 100 volts at its peak and **V2** is 75 volts at its peak, as set by their parameters. Both voltages attain zero volts every 8.35 milliseconds, as determined by their frequency.

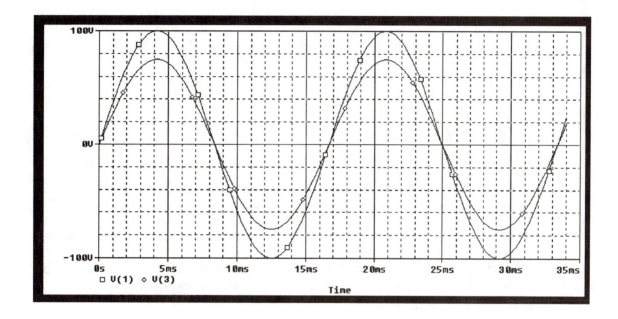

The two **PROBE** plots show all the currents through and the voltages across all resistors. All of them are sine waves with zero phase. However, both the voltage V(2,3) and the current I(R2) have inverted peak values compared to the other currents and voltages. Why is this so? The current I(R2) is driven counterclockwise through **R2** by the voltage source **V2**. Thus, it enters the negative terminal of **R2** as dictated by the syntax of the **PSpice** program. The reader is referred to Chapter 1 for a discussion of the polarity of **PSpice** components. Voltage V(2,3) is negative because of the direction of the flow of current I(R2). The power to **R2**, which is equal to the product of V(2,3)*I(R2), is positive. This means that **R2** is receiving power as it should. Equally, the power to **R1** and **R3** is also positive as will be shown.

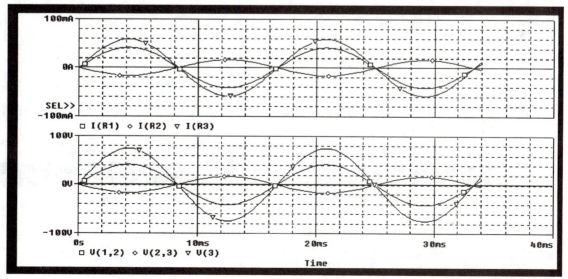

The power flow for Figure 4.03 is shown next.

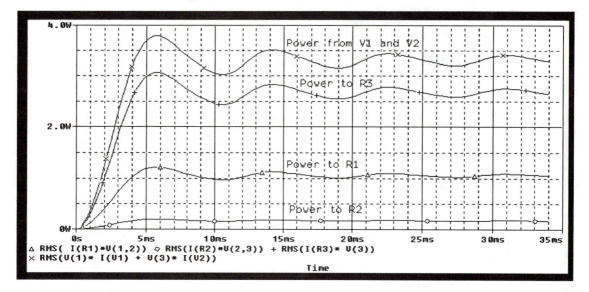

The **PROBE** plot shows that the power to all resistors is positive. Let us verify that the power delivered by the voltage sources **V1** and **V2** is equal to the power received by the three resistors. We write the following equation:

power from the sources **V1** and **V2** = power to resistors **R1**, **R2** and **R3**

Hence:

$$RMS(V(1)*I(V1) + V(3)*I(V2)) = RMS(I(R1)*V(1,2) + I(R2)*V(2,3) + I(R3)*V(2))$$

The **PROBE** plots of these expressions is shown next.

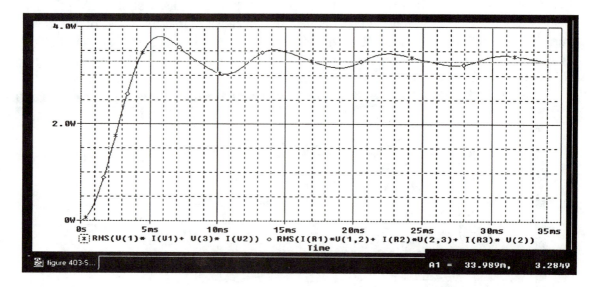

The plots of the power from the voltage sources **V1** and **V2** and to the resistors **R1**, **R2** and **R3** overlap completely. This proves the equivalence of the two expressions. The total power delivered and received is shown to be 3.2849 watts by cursor **A1**. The reader is encouraged to verify that result.

TIME DELAY AND PHASE SHIFT IN SINE WAVES

What's new?
1. The use of **TD** in sine waves
2. The **Arrow** function

We next study sine waves that do not all cross the time axis at the same time. Such sine waves, by previous definition, are not in phase. Let us analyze the circuit in Figure 4.04.

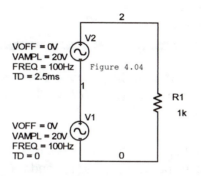

The **TD** (time delay) parameters for **V1** and **V2** must be set from the **Property Editor** sheet in the same manner as was done in previous examples. Enter 2.5 ms into **TD** box. Click on **Display** and **Apply**. The legend **TD = 2.5** ms will appear next to symbol for **V2**. Set the **TD** for **V1** of zero milliseconds in the same manner.

Source Package	TD	Value	VAMPL	VOFF
VSIN	2.5ms	VSIN	20V	0V

A **transient analysis** of 20 milliseconds will plot about two cycles of all the voltage sine waves. The results are shown.

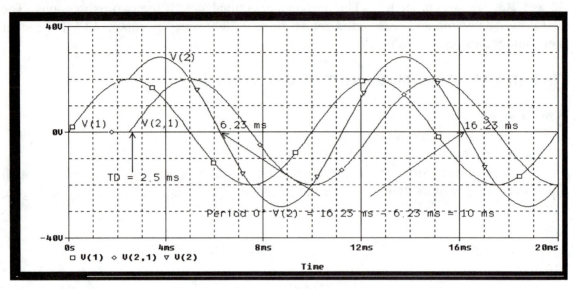

On this plot, V(1) is equal to the voltage of the source **V1**. V(2,1) is equal to the voltage of source **V2**. The nodal voltage V(2) is the voltage across the resistor **R1**. It is also equal to **V(2,1) + V(1).**

V(2,1) is delayed by 2.5 milliseconds, as called for by its **TD**. V(2) is a sine wave that peaks midway between the peaks of V(1) and V(2,1). Its peak value of about 28 volts is not equal to the arithmetic sum of the peaks of **V(1)** and **V(2,1)**. That sum would have been 40 volts. This is an important point to remember. The period of V(2) is equal to 10 milliseconds, as shown. Its frequency of 100 Hz is the same as that of the voltage sources **V1** and **V2**.

To obtain the arrows as shown on the **PROBE** plot, if such is desired:
1. With **PROBE** running, click on **Plot.**
2. Click on **Label.**
3. Click on **Arrow.** The cursor changes to pencil form.
4. Position the tail of the arrow where desired and click.
5. Move the cursor to the desired position of the head of the arrow and click. The complete arrow will appear.

The Relationship between Time Delay and Phase Shift

In engineering usage**,** a time shift between sine waves is traditionally expressed not in units of time but in degrees. We shall explore their relationship. At the outset let us note that a time delay is a physical event, whereas a phase shift expressed in degrees is a mathematical artifact.

A sine wave goes through a complete cycle in 360° For instance, a conductor rotating in a magnetic filed rotates through 360° in producing one sine wave. This journey also takes a finite amount of time, the period of the sine wave. We can express the relationship between phase shift and time delay by the following proportion:

$$\frac{Phaseshift}{360} = \frac{TD}{Period}$$

Solving for Phase shift, we have: $phaseshift = 360 * \frac{TD}{Period} = 360 * \frac{2.5ms}{10ms} = 90°$.

Thus, the **TD** of 2.5 milliseconds for V(2,1) results in a 90° phase lag relative to V(1). This can be seen in the above **PROBE** plot. A negative **TD** of 2.5 milliseconds would have moved V(2,1) 90° ahead of V(1).

MULTIPLE PHASE SYSTEMS
What's new?
1. Specifying Phase(**PH**)

A Two-Phase System

Modern electrical power systems are multiphase systems. Of particular prominence is the three-phase system. We shall introduce this subject by first studying the two phase system of Figure 4.05 and then proceed to the analysis of a three-phase system.

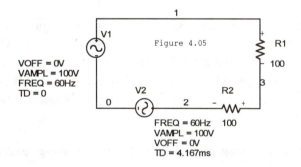

Voltage sources **V1** and **V2** have their negative terminals connected to ground, the node 0. Resistor **R2** has been rotated 180° so that its positive terminal at node **3** is connected to the negative terminal of **R1** at the same node. In this way, the current I(R1) or I(R2) will always enter or exit resistor terminals having the same polarity. If this is not observed, an erroneous phase shift will appear in the data, one not based on physical fact but upon the syntax of the **PSpice** program. Beware!

V2 has a **TD** of 4.167 milliseconds. This corresponds to a phase shift of 90° relative to **V1**. A **transient analysis** of 34 milliseconds produced the traces shown. Nodal voltages V(1) and V(2) are the source voltages **V1** and **V2.** Their amplitudes are 100 volts. V(1,2) has a peak of 141 volts when V(1,0) and V(0,2) are of equal amplitude but have opposite polarity. This happens, by way of example, for a positive peak at $t = 18.710$ milliseconds and for a negative peak at $t = 27.043$ milliseconds. The time difference between the peaks of V(1,2) is 8.333 milliseconds. This is equal to half the period of V(1,2); hence its frequency is 100 Hz. V(1,2) crosses the time axis when V(1) and V(2) are of equal amplitude and have the same sign.

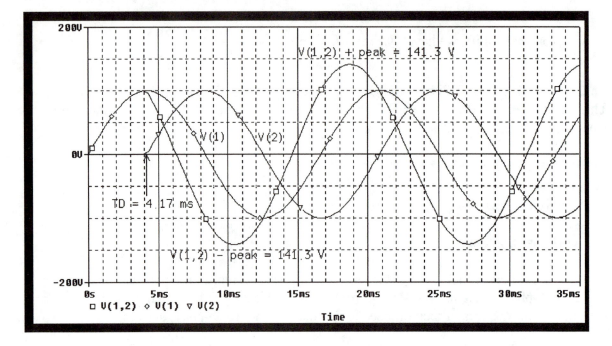

The current through **R1** and **R2** is shown next. Figure 4.05 is a series circuit; therefore, only one current can be flowing. It follows that currents I(R1) and I(R2) are identical; hence they do not have a relative time delay or phase shift. The amplitude of the current is 707.335 milliamps. This, when multiplied by **(R1+R2),** is equal to 141 volts. That voltage is equal to the voltage V(1,2) as previously determined. All this is shown on the above **PROBE** plot.

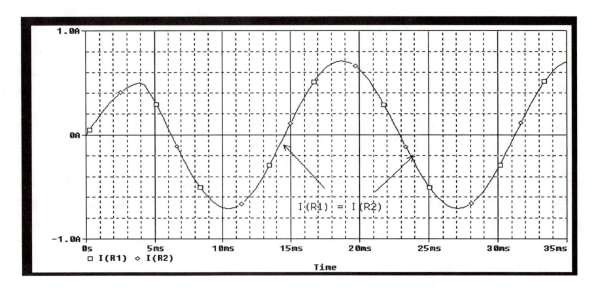

The next **PROBE** plot shows that the voltages across **R1** and **R2** are in phase and their sum is 141 volts. That is equal to the voltage V(1,2), as it must be to satisfy Kirchhoff's voltage law.

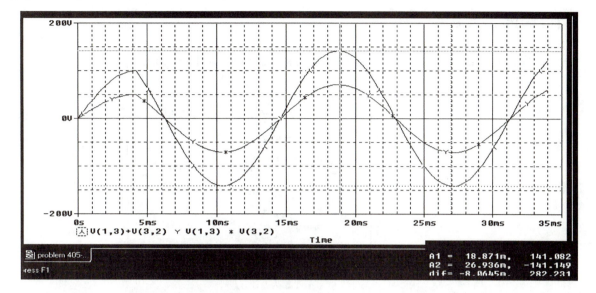

The next **PROBE** plots shows the current flowing through the two voltage sources **V1** and **V2**. The current I(V1) has a relative phase shift of 180° relative to the current I(V2). This is because I(V1) leaves the positive terminal of **V1** while I(V2) enters the positive terminal of **V2**.   The syntax of **PSpice** defines the former current as negative and the latter as positive. When the circuit current flows counterclockwise, the sign of the two source currents is reversed. The reader is reminded again that there is only one circuit current flowing in the circuit.

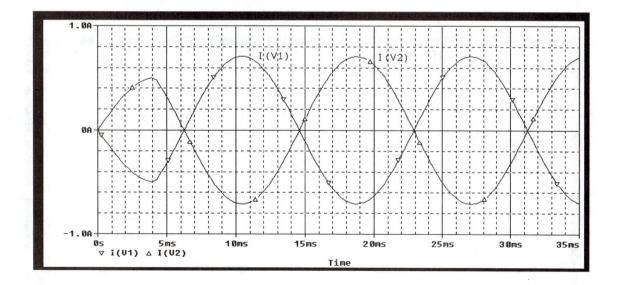

A Three-Phase System

We shall next analyze the three-phase system shown in Figure 4.06.

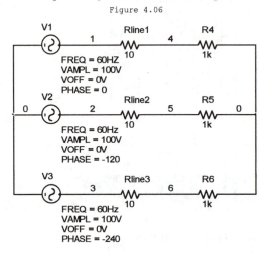

Figure 4.06

In the above circuit, the phase shifts of **V1**, **V2** and **V3** were specified rather than their **TD**. The procedure is the same in both cases, with the difference that the desired phase shift, such as –120 degrees, is entered in the **PH** box in the **Property Editor** sheet. A **Transient analysis** of 35 milliseconds was run. We start by examining the three traces of the three voltage sources **V1**, **V2** and **V3**. Their nodal voltages are shown next.

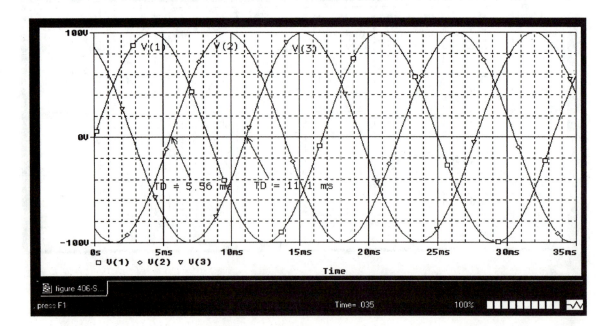

All three voltages have peaks of plus or minus 100 V. **V(2)** is delayed relative to **V(1)** by 5.56 milliseconds, which corresponds to a phase shift of -120°. **V(3)** is delayed relative to **V(1)** by 11.1 milliseconds, which corresponds to a phase shift of -240°. These voltages are defined as the phase voltages of the sources.

The voltage across each resistor is shown next.

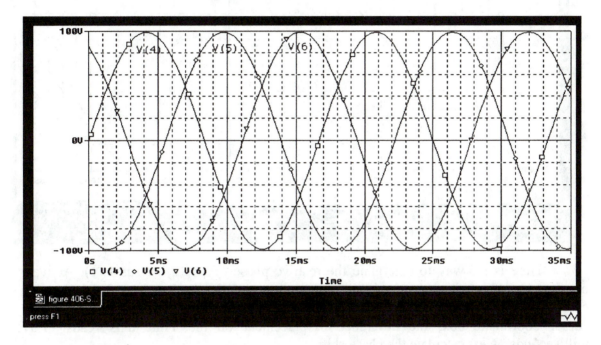

Comparing these voltages with the source voltages we observe that **V(4)**, **V(5)** and **V(6)** are identical to **V(1)**, **V(2)** and **V(3)**. This becomes apparent when we study the circuit in Figure 4.06. **V1** is directly connected to **R4**, neglecting the small voltage drop across **Rline1**. The line resistors were added to better identify the nodal voltages of the resistors **R1**, **R2** and **R3**. **V(4)**, **V(5) and V(6)** have a 120° phase shift relative to each other, as did **V(1)**, **V(2) and V(3)**. We plot the voltages V(1,2), V(1), V(2) and V(3) to compare them.

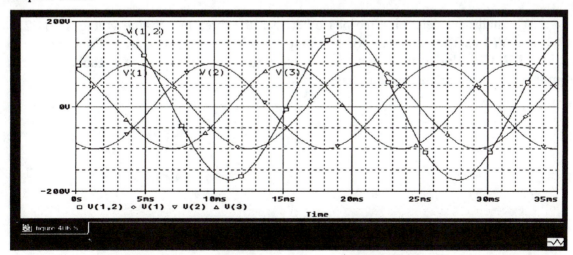

The voltage **V(1,2)** is defined as a line-to-line voltage. Its peak amplitudes are plus or minus 173 V. These peaks are 8.3871 ms apart, since the frequency is 60 Hz. We next determine the phase shift of **V(1,2)** relative to the phase voltages **V(1)**, **V(2)** and **V(3)**.

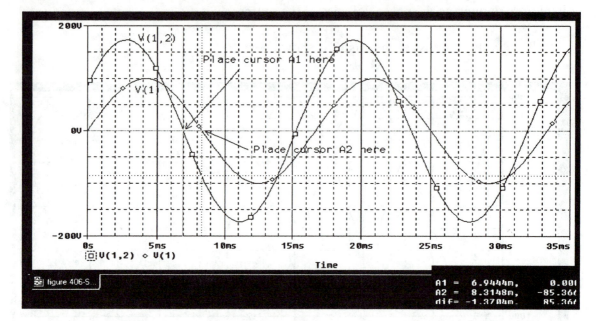

Here is the way to determine the relative phase of **V(1,2)** versus **V(1)**. Activate cursor **A1** and place it as shown above. Activate cursor **A2** and place it as shown. In each case, the vertical cross hair of the cursors is at the zero crossing of **V(1,2)** and **V(1)**. In the cursor coordinate box, the difference in placement on the **Time** axis reads −1.37 milliseconds. Next calculate the phase shift:

$$Phaseshift = 360 * \frac{TD}{Period} = 360 * \frac{1.37ms}{16.67ms} = 30°$$

Thus, the line-to-line voltage **V(1,2)** leads the phase voltage **V(1)** by 30°. In the same manner, the phase shift of the other two line-to-line voltages relative to the phase voltages can be determined. This is left as an exercise for the reader.

We shall next determine the phase between the three line-to-line voltages **V(1,2)**, **V(2,3)** and **V(3,1)**. The procedure for obtaining them is the same as above. First, we shall obtain a **PROBE** plot of them and then measure the time delay between each of them. From that delay, we calculate their phase shifts.

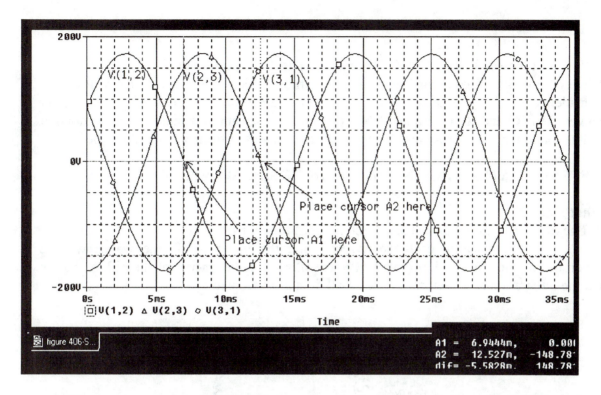

Placing cursors **A1** and **A2** as shown, the relative **TD** of **V(1,2)** versus **V(2,3)**   is −5.582 milliseconds. From this we calculate that the line voltage **V(2,3)** lags 120° behind line voltage **V(1,2).** Using the same procedure to determine the relative phase shift between **V(1,2)** and **V(3,1)**, we find that **V(3,1)** lags **V(1,2)** by 240°. A system in which all voltages have the same amplitude and the same relative phase angle is defined as a balanced system.

The power delivered by **V1** to **R4** is shown next.

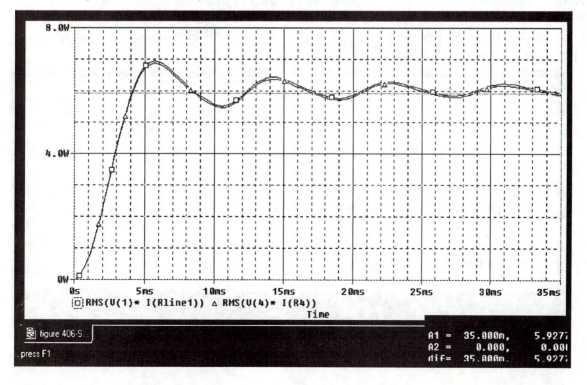

That power is equal to 5.927 watts. The first expression is the **RMS** power delivered by **V1**. The second expression is the **RMS** power delivered to **R4**. The coincidence of the two traces proves the equivalency of these powers. The power delivered to **R5** and **R6** by **V2** and **V3** is obtained in the same manner. If such is done, it will be found that **R5** and **R6** also receive about 6 watts each from **V2** and **V3** respectively.

DIFFERENTIATION AND INTEGRATION OF SINE WAVES

What's new?

1. Use of the **D**(differentiation) and **S**(integration) functions

The determination of the derivatives and the integrals of sine waves will be important in much of our future work. **PSpice** allows for the tracing of the differentiation and the integration of waveforms, as we shall demonstrate next.

Figure 4.07

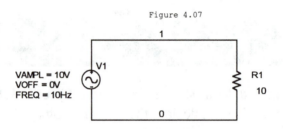

Figure 4.07 shows a voltage sine wave with the given parameters applied across **R1**. Our objective is to find the derivative and the integral of that voltage. First we shall obtain the derivative. A **transient analysis** of 200 milliseconds produced the traces shown. **V1** is the source voltage, and **D(V(1))** is the derivative of that voltage. The **D**(differentiator) operator is in **the Functions or Macros** dialog box. To open it, click on **Trace**, click on **Add Trace**, scroll to **D** and click on it to select it. The symbol **D( )** will appear in the **Trace Expression** box. Click on **V(1);** it will be entered into the **Trace Expression** box, as shown below.

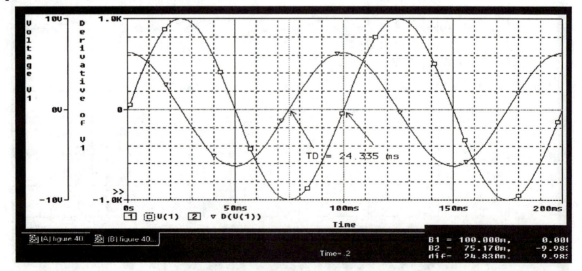

The derivative of **V(1)** is a cosine wave. It starts at its peak value at $t = 0$ seconds and it leads **V(1)** by 24.83 milliseconds. This computes to a 90° phase shift. We recall from elementary calculus that

$$\frac{d}{dt}(\sin u) = \cos u \frac{du}{dt} \quad \text{where} \quad u = \omega t, \quad \text{therefore} \quad \frac{du}{dt} = \omega$$

By definition, ω is the radian frequency of **V1**, hence $\omega = 2\pi f = 2*3.14*10 = 2.8$ radians/second. Plugging this value and the amplitude of **V1** of 10 V in the above derivative equation, we obtain:

$$\frac{d}{dt}(10\sin 62.8t) = 10*62.8*\cos 62.8t \quad \text{volts}$$

This tells us that the peak positive amplitude of the derivative is equal to 628 V. The **PROBE** plot shows that it is!

We next determine the integral of **V1**. The procedure is the same as for the determination of the derivative with the exception that the **S**(integrator) symbol is selected in the **Function and Macros** scroll box. A **transient analysis** of 200 milliseconds produced the traces shown.

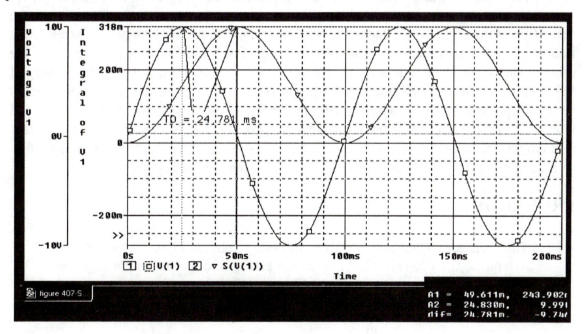

The integral of V(1) is the trace of S(V(1)). It is a cosine wave with a positive offset of 159 millivolts. This offset is the constant of integration, which, when differentiated, is equal to zero volts. The integral of a function is proportional to the area under the function. This can be seen from the preceding plot. The area of V(1) reaches a maximum at t = 50 milliseconds. At that instance, S(V(1)) is at its maximum of 318 millivolts. For the interval 50 milliseconds < t < 100 milliseconds, the net area of V(1) declines till at t = 100 milliseconds, it is zero volts. Therefore, S(V(1)) has a zero value. Thereafter, the process repeats itself. From calculus we have:

$$\int \sin u du = -\cos u + C$$

where u = ωt, since $\dfrac{du}{dt} = \omega$, therefore du = ωdt and ω = 62.8 radians/second, as before.

Substituting the last equation and adding the amplitude of **V(1)** of 10 V results in the following equation depicted by **S(V(1)).**

$$10\int \sin \omega t dt = -\frac{10}{62.8}\cos \omega t dt + C = -.159\cos \omega t + C$$

Measuring the distance between adjacent peaks of the voltages V(1) and its integral S(V(1)) shows that the latter lags the former by 24.781 milliseconds. This calculates to a phase delay of 90°.

From the foregoing operations of differentiation and integration of a sine wave, we found that these operations resulted in another sine wave. In the case of differentiation the derivative of the original sine wave was a sine wave with a changed amplitude and a phase lead of 90° relative to the original sine wave. In the case of integration, the integral of the original sine wave resulted in a sine wave with a changed amplitude and a phase delay of 90° relative to the original sine wave. In Chapter 2 we noticed the integro-differential relationships that existed between currents and voltages relative to capacitors and inductors. We will find that when a voltage sine wave is applied to a capacitor, the current through the capacitor will lead that voltage by 90°. When a voltage sine wave is applied to an inductor, the current through the inductor will lag that voltage by 90°.

PROBLEMS

4.1   The sinusoidal voltage source **V1** has a peak amplitude of 100 volts and
       a frequency of 1000 Hz. Find the voltage across, and the current through   each
       resistor. Prove that the product   AVG(V(4)*I(R6)) is equal to
       RMS(V(4))*RMS(I(R6)).

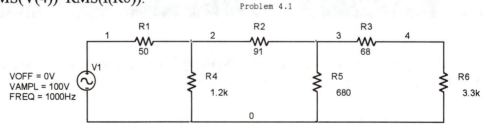
Problem 4.1

4.2   The sinusoidal voltage source has a peak value of 100 volts and a
       frequency of 100 Hz. A 50 volt dc source has been added to the circuit,
       as shown. For two cycles of the sinusoidal voltage source:
       (a). Find all absolute and relative voltages
       (b). Determine the current through each resistor.
       (c). Compare the dc offsets of the absolute voltages with those
            of the relative voltages. Explain why they differ.
       (d). Find the instant, the average and the **RMS** power delivered
            to **R6**.

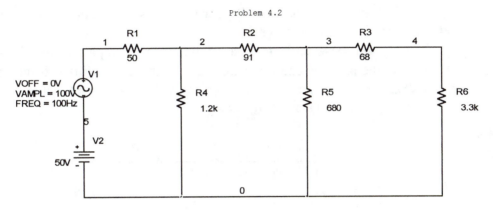

Problem 4.2

4.3   Obtain the circuit voltages and the circuit currents. Find the **RMS** power to all
       resistors and compare it with the **RMS** power delivered by **V1** and **V2**. Which of
       these two voltage sources delivers the larger amount of power?

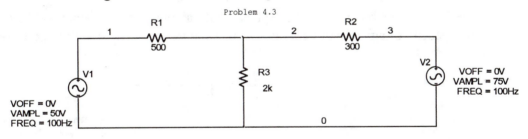

Problem 4.3

4.4   Make a **PROBE** plot of the source voltages **V1** and **V2**. From it, verify their
       amplitudes and relative phase shifts. Find the nodal voltages V(1,3) and **V(3).**
       What are their amplitudes and relative phase shifts. Verify Kirchhoff's voltage law
       for this circuit. Obtain the **RMS** power delivered to **R1** and **R2**. Compare that with
       the total **RMS** power delivered by **V1** and **V2**.

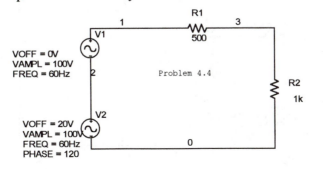

Problem 4.4

4.5   The two voltage sources **V1** and **V2** are used to supply electrical power through **Rline1** and **Rline2** to **Rload**. The initial phase shift of **V1** and **V2** are both 0°. For this condition find the voltage across and the current through **Rload** and the **RMS** power received by it.

Next, change the phase of **V2** to a phase shift of -30° and repeat the analysis. Which source is contributing power to **Rload?** Which source is receiving power?

Restore the phase shift of **V2** to 0° and change the phase shift of **V1** to minus 30°. Repeat the analysis. Which source is contributing power to **Rload?** Which source is receiving power?

Keep the phase shift of **V2** at 0°, but change the phase shift of **V1** to minus 60°. Repeat the analysis and compare the present data with that from the previous runs. What generalizations can the reader draw from the results obtained? Is there a practical use for such a system?

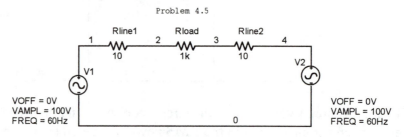

Problem 4.5

4.6   The 50 V dc source **V1** is to be replaced with a sine wave generator of 60 Hz. Find the needed voltage of that generator so that the same power is delivered to **Rload**.

Problem 4.6

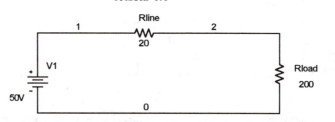

4.7   Find the voltages across **R1** and **R2. Find** the **RMS** power delivered by **V1** and **V2** to resistors **R1** and **R2** over two cycles of **V1**. Is it legitimate to obtain that **RMS** power even though the voltages **V(1,3)** and **V(3)** are not sine waves ?

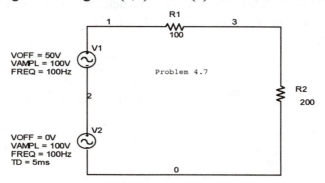

4.8   Find the instantaneous current through, and voltage across **R4**. What is the **RMS** power delivered to **R4** for three cycles of **V1.** Explain the relative phase shift between V(1) and V(3).

Problem 4.8

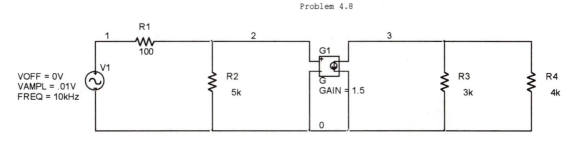

4.9   Perform a **transient analysis** of 20 milliseconds.
        (a). Find the voltage V(2) and its phase shift relative to V(1).
        (b). Find the average voltage AVG(V(2)).
        (c). Find the **RMS** voltage of V(2).
        (d). Compare AVG(V(2)) with RMS(V(2)).
        (e). Obtain the average **(AVG)** power to **Reload.**
        (f). Obtain the **RMS** power to **Rload**.
        (g). Compare the **AVG** power to Rload with its **RMS** power.
        (h). What is the **RMS** power delivered by voltage source **V1?**
        (i). What is the **RMS** power delivered by voltage source **V2**?
        (j). Is there a difference in the average power **AVG** delivered by **V2** compared to the **RMS** power delivered by that source?

Problem 4.9

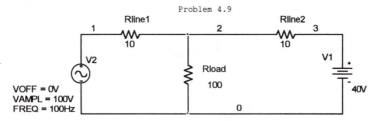

4.10   The circuit shown has a voltage source **V2** with a 50 V offset. Run a 40 millisecond **transient analysis.**
        (a). Obtain a **PROBE** plot of the traces of V(1) and V(2), compare them.
        (b). Obtain the average voltage of V(1) and V(2) after 40 milliseconds.
        (c). What is the difference in these averages and why?
        (d). Obtain the **RMS** values of V(1) and V(2), and compare them.
        (e). Plot V(3), its **AVG** and **RMS** values on one **PROBE** plot.
        (f). Compare these traces and explain the differences.
        (g). Obtain the **RMS** power delivered by **V1** and **V2** at t = 40 milliseconds.
        (h). Which of the two sources delivers more power and why?
        (i). Use two **Y-plots** to get the traces of V(3**)** and I(R3).

(j). Compare their amplitudes and relate them to the value of **R3.**
(k). Are V(3) and I(R3) in phase?
(l). Obtain the value of the **AVG** and the **RMS** power to **R3** at $t = 40$ milliseconds
(m). Which of these is higher and why?

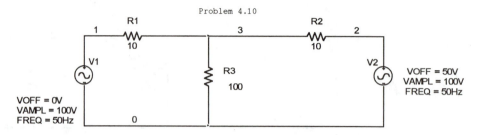

Problem 4.10

4.11   Perform a **transient analysis** of 20 milliseconds duration.
(a). On one **PROBE** plot, get the traces of **I1**, I(R1), I(R2) and I(R3).
(b). Why is there a 180° phase shift between **I1** and the resistor currents?
(c). Verify that the sum of the resistor currents is equal to the source current **I1**.
(d). What is the **RMS** power delivered by the current source **I1.**
(e). Find the **RMS** power to each resistor. Use one **PROBE** plot.
(f). Which resistor receives the most and which receives the least power?
(g). Verify that the sum of the **RMS** power received by the resistors is equal to the power delivered by the current source **I1**.
(h). What is the value of that power?

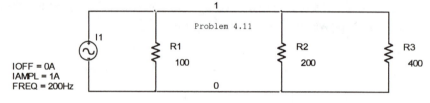

Problem 4.11

4.12   This circuit has a current source **I1** and a voltage source **V1** connected to **R1** as shown. Perform a **Transient analysis** for 2 milliseconds.
(a). On one **PROBE** plot, obtain the traces of V(1) and V(1,2).
(b). Add a second Y-axis and obtain the trace of V(1) and V(2).
(c). Are all three voltages in phase?
(d). What are the amplitudes of V(1) and V(1,2)?
(e). What is the relative phase, if any, between I(I1) and V(1)?
(f). Obtain the **RMS** power to **R1**.
(g). Change the phase shift of **V1** to -30° and run the **PSpice** analysis again.
(h). Answer parts (a) through (f) again.
(i). Compare the voltage and the power levels of the two runs.

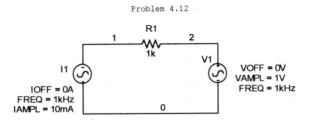

4.13   The circuit has three voltage sources, **V1**, **V2** and **V3** connected across resistor **R1** as shown. Perform a **transient analysis** for 20 milliseconds.

(a). On one **PROBE** plot, plot the traces of V(1,2), V(2,3) and V(3)

(b). What is their relative phase shift?

(c). Plot the voltage V(1).

(d). Verify that the sum of the source voltages is equal to the resistor voltage V(1).

(e). Make one **PROBE** plot of the current I(R1).

(f). Use a second **Y-Axis**, and plot the three source voltages.

(g). From a comparison of these plots, which of the voltages is leading I(R1), which one is in phase with I(R1) and which voltage is lagging I(R1)? Explain your findings.

(h). Find the total **RMS** power delivered to **R1** and the contribution of each voltage source to that power.

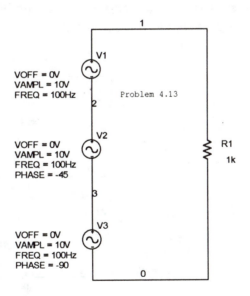

4.14   The circuit shown has two voltage sources **V1** and **V2**, with their indicated phase shift. Run a **transient analysis of** 4 milliseconds.

(a). Verify from a **PROBE** plot, the relative phase shift between **V1** and **V2**.

(b). On one **PROBE** plot, get the traces of V(1), V(2) and V(3).

(c). From that plot, get their amplitudes and the relative **TDs** (time delays).

(d). Convert these **TDs** into phase shifts.

(e). Get the **RMS** power to **R1**, **R2** and **R3**.

(f). Which of them receives the smallest and which receives the largest amount of power?

(g). Get the **RMS** power delivered by **V1** and **V2**.

(h). Verify that the **RMS** power delivered by the sources **V1** and **V2** is equal to the **RMS** power received by **R1**, **R2** and **R3**.

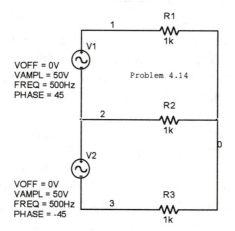

4.15   This circuit is essentially the same as in the previous problem. However, the nodal assignment has changed. Answer all questions as in the previous problem and determine if there has been any change in the amplitudes of the circuit voltages and the **RMS** power delivered to **R1**, **R2** and **R3**.

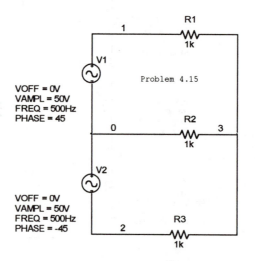

4.16   The circuit in this problem is a balanced three-phase system. The elements connected in the manner shown form a **Y-Y** connected system. In it, voltages **V(1)**, **V(2)** and **V(3)** are referred to as the phase voltages. The voltages **V(1,2)**, **V(2,3)** and **V(3,1)** are referred to as the line voltages. The nodes connecting the load resistors have been labeled with capital letter to help in their identification. Such a procedure is legitimate in **PSpice**.   Perform a **transient analysis** of 34 milliseconds.

(a). Obtain a **PROBE** plot of the three phase voltages V(A), V(B) and V(C)

(b). Are their amplitudes as specified?

(c). Measure their **TD**s (time delays) and convert them to phase shifts.

(d). Make a **PROBE** plot of the line voltages V(1,2), V(2,3) and V(3,1). Note that the nodal sequence begins with node 1 and ends in node 1.

(e). Make a **PROBE** plot of V(1,2)/V(A) and of I(V1)/I(Rload1). Remember these ratios.

(f). Make a **PROBE** plot of V(1) + V(2) + V(3).

(g). Make a **PROBE** plot of I(Rload1) + I(Rload2) + I(Rload3).

(h). Find the **RMS** power delivered to **RLoad1, Rload2** and **Rload3.**

(i). The **RMS** power delivered by each voltage source.

(j). Compare the power delivered by the voltage sources with that received by the load resistors. Are they the same ?

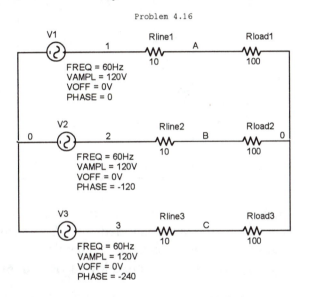

Problem 4.16

4.17   The circuit shown is a balanced Δ-Δ (delta) system. The various voltages and currents are classified as follows:

Line voltages: V(1), V(2), V(3)   Load voltages: **V(A), V(B), V(C)**

Source currents: I(V1), I(V2), I(V3)   Load currents: I(R1), I(R2), I(R3)

Line currents: I(RlineA), I(RlineB), I(RlineC)

In the circuit, the **Rbreak** resistors were added because **PSpice** will return a loop error message before executing a run and advise the user to break the loop consisting of the voltage sources with a small resistor in series with the sources. This was done as shown. Perform a **Transient analysis** of 34 milliseconds and use the **PROBE** plots.

(a) Obtain the amplitudes and relative phase shifts of the line voltages.

(b). Find the amplitudes and relative phase shifts of the load voltages.

(c). Get the ratio of a line voltage divided by a load voltage.

(d). Get the phase shift of a line voltage relative to a load voltage.

(e). Determine the amplitudes and relative phase shifts of the source currents.

(f). Find the amplitude and relative phase shifts of the line currents.

(g). Find the amplitude and the relative phase shifts of the load currents.

(h). Get the ratio of a line current divided by a load current.

(i). Get the ratio of a line current divided by a source current.

(j). Get the phase shift of a line current relative to a load current.

(k). Get the **RMS** power received by each load resistor.

(l). Get the **RMS** power delivered from each voltage source.

(m). Compare the results of the previous two requests.

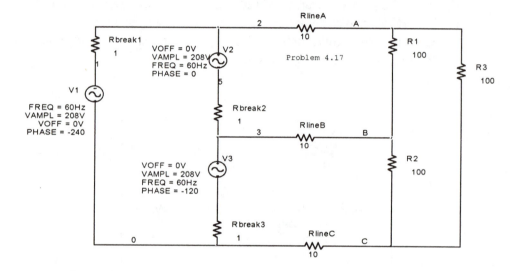

4.18   The system shown is a Δ-Y connected one. While the voltage sources are in the delta configuration, the load resistors are connected as a Y. Perform a **transient analysis** as before and answer all questions as in Problem 4.17. In the labeling of the load resistors, no spaces between letters are allowed. **Rload A** is not OK, but **RloadA** is.

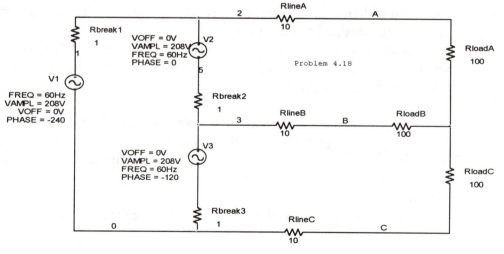

4.19   The system shown is a Y-Δ connected one. In this case, the sources are in Y
configuration, while the load resistors are connected in Δ. Perform a **transient
analysis** of 34 milliseconds and answer all questions as in Problem 4.17.

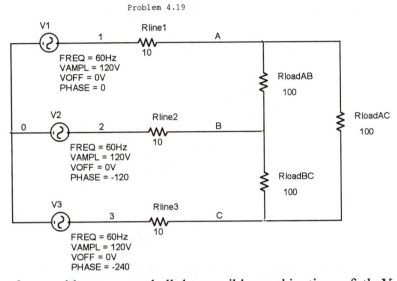

Problem 4.19

4.20   The previous four problems covered all the possible combinations of the Y-Δ
connections between voltage sources and resistive loads in a three phase system.
Compare the ratio of the line voltage to the load voltages and the line currents to
the load currents for the four systems covered in Problem 4.16 through Problem
4.19 What conclusions can be drawn?

4.21   The circuit has two voltage sources, **V1** and **V2**, connected as shown. Both have an
amplitude of 1 volt and zero relative phase. **V2** has an offset voltage of 1 volt. For
this circuit:
(a). Find the voltage V(1) and its derivative. Is the latter a cosine wave?
(b). Verify the amplitude of D(V1)) analytically.
(c). Find the current I(R1) and its phase shift relative to V(1).
(d). From plot of D(I(R1)) and D(V(1)), compare the two relative to amplitude and
phase shift.
(e). Is D(V(1,2)) + D(V(2)) = D(V(1)) ?

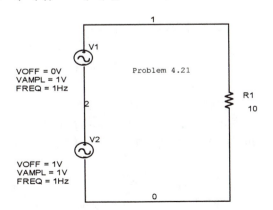

Problem 4.21

4.22   Repeat Problem 4.21 for the circuit shown. Note that the offset voltage on **V2** has been removed. **V2** now has a relative phase shift of 90°. What is the difference in the present data compared to the previous problem? In particular, is D(V(1,2)) + D(V(2)) = D(V(1)) ?

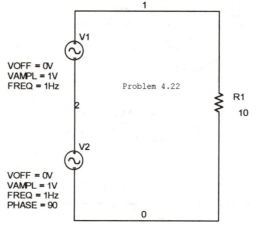

4.23   The circuit has a sinusoidal voltage source **V1** applied to it. That source has a phase shift of 180°. What is the slope of this voltage at t = zero seconds? Find the integral of that voltage and classify it as either a sine or cosine wave. Is there a phase shift between I(R1) and V(1) ? Is there a phase shift between S(I(R1)) and S(V(1))? It is best to use two **Y-plots** to answer the last two questions.

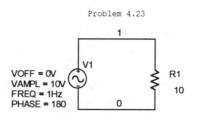

4.24   For this circuit, perform a **transient analysis** for 200 milliseconds.
(a). Obtain a **PROBE** plot of V(1), V(2) and V(3).
(b). From this data, determine the relative phase shifts and **TDs** of these three voltages.
(c). Using two **Y-plots**, on one get the trace of V(3) and on the other the trace of S(V(3)).
(d). What are their amplitudes and what is the relative phase shift between these two voltages?
(e). Use two **Y-plots**, get the traces V(3) and D(V(3)).
(f). What is the amplitude of D(V(3)) and its phase shift relative to V(3)?
(g). Use two **Y-plots**, on one, get the trace of S(V(3)) and on the other the trace of D(V(3)).

(h). What is the relative phase shift between these two?

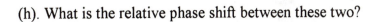

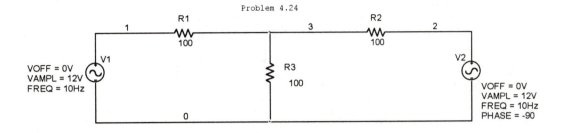

STEADY-STATE SINUSOIDAL RESPONSES OF RC, RL AND RLC CIRCUITS

SINUSOIDAL RESPONSES OF CAPACITORS AND INDUCTORS

In this chapter, we expand the application of sinusoidal currents and voltages by including capacitors and inductors in addition to resistors in our circuits. The values of the circuit elements are chosen in such a way that any transients decay quickly when compared with the steady-state response of our circuits. The following voltage/current and phase relations hold for a resistor, a capacitor and an inductor, respectively.

For a resistor, the ratio of voltage/current is equal to the resistance of the resistor. Its symbol is $\mathbf{Z_R}$, or most often simply $\mathbf{R}$. Its unit is the ohm (Ω). Its voltage and current are in phase. The reciprocal of that ratio, current /voltage, is defined as the conductance of the resistor. Its symbol is $\mathbf{G}$ and its unit is the siemens (S). Its current and voltage are in phase.

Symbolically: $\quad \mathbf{Z_R} = R\angle 0°$ ohms; $\quad \mathbf{G_R} = \dfrac{1}{R}\angle 0°$ siemens

For a capacitor, the ratio of voltage/current is inversely proportional to the product of the applied radian frequency and its capacitance. The amplitude of this ratio is defined as the reactance of the capacitor. Its symbol is $\mathbf{X_C}$ and its unit is the ohm (Ω). The voltage across a capacitor lags the current through it by a quarter of a wavelength of the period of the applied frequency. This is equal to a - 90° phase shift. The reactance together with the phase shift is defined as the impedance $\mathbf{Z_C}$ of the capacitor.

The amplitude of the reciprocal of that ratio, current/voltage, is defined as the susceptance of the capacitor. Its symbol is $\mathbf{B_C}$ and its unit is the siemens (S). The current across a capacitor leads its voltage by a quarter of a wavelength of the period of the applied frequency. This is equal to a plus 90° phase shift. The susceptance together with the phase shift is defined as the admittance $\mathbf{Y_C}$ of the capacitor.

Symbolically: $\quad \mathbf{Z_C} = \dfrac{1}{\omega C}\angle -90°$ ohms; $\quad \mathbf{Y_C} = \omega C\angle 90°$ siemens

For an inductor, the ratio of its voltage/current is directly proportional to the product of the applied radian frequency and its inductance. The amplitude of this ratio is defined as the reactance of the inductor. Its symbol is $\mathbf{X_L}$ and its unit is the ohm (Ω). The voltage across an inductor leads the current through it by a quarter of a wave length of the period of the applied frequency. This is equal to a + 90° phase shift. The reactance together with the phase shift is defined as the impedance $\mathbf{Z_L}$ of the inductor.

The amplitude of the reciprocal of that ratio, current/voltage, is defined as the susceptance of the inductor. Its symbol is $\mathbf{B_L}$ and its unit is the siemens (S). The current across an inductor lags its voltage by a quarter of a wavelength of the period of the applied frequency. This is equal to a minus 90° phase shift. The susceptance together with the phase shift is defined as the admittance $\mathbf{Y_L}$ of the inductor.

Symbolically: $\mathbf{Z_L} = \omega L \angle 90°$ ohms; $\mathbf{Y_L} = \dfrac{1}{\omega L} \angle \text{-}90°$ siemens

The amplitude and phase relations of these three circuit elements are a direct consequence of the integro-differential relations that exist between their respective currents and voltages. In each case, the angle between the voltage and the current for each of these elements is defined as the impedance angle of their impedance or the admittance angle of their admittance.

In circuits containing resistors and capacitors, the impedance angle can be between -90° and 0° and the admittance angle between 0° and 90°. Therefore, the input voltage will always lag the circuit current. In circuits containing resistors and inductors, the impedance angle can be between 0° and 90° and the admittance angle between -90° and 0°. Therefore, the input voltage will always lead the circuit current.

In circuits containing resistors, capacitors and inductors, if the capacitive reactance is equal to the inductive reactance, the impedance angle is zero and the total impedance of the circuit is equal to the resistance of the circuit. Input voltage and circuit current are in phase. If the capacitive reactance is greater than the inductive reactance, the impedance angle will be less than 0°. The input voltage will lag the circuit current. If the inductive reactance is greater than the capacitive reactance, the impedance angle will be greater than 0°. The input voltage will lead the circuit current. The reader is encouraged to perform the analysis in this paragraph in terms of the admittances and their associated angles.

Sinusoidal Voltage Applied to a Capacitor

Figure 5.01 shows a circuit in which a sinusoidal voltage source **V1** is applied across a capacitor. We shall perform a 2 millisecond **transient analysis** to get the circuit current. The analysis produced the **PROBE** plot shown.

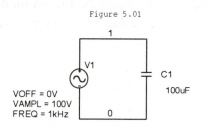

Figure 5.01

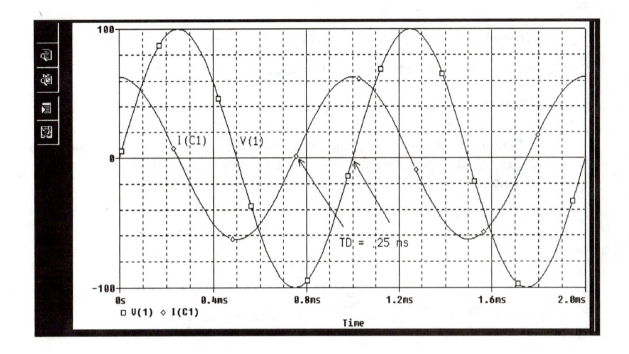

The **PROBE** plot shows V(1) with an amplitude of 100 volts lagging the current I(C1) with amplitude 62.8 amps by .25 milliseconds. This computes to a 90° phase lag, as was predicted. To obtain that **TD**, proceed as follows: Activate cursor **A1** and place it at a convenient zero crossing of V(1). Activate cursor **A2** and place it at the peak of V(1) before the selected zero crossing. This corresponds to a zero crossing of I(C1). The **TD** indicated in the cursor box of 251.511 μs is the **TD(time delay)** between V(1) and I(C1).

The reader will recall that the current/voltage relations for a capacitor are governed by the following equations:

$$i(t) = C \frac{dv}{dt} \text{ amperes } \text{ and } v(t) = \frac{1}{C} \int i(t) dt \text{ volts}$$

The applied voltage to the capacitor of Figure 5.01 is:

$v(t) = 100 \sin \omega t$ volts, where $\omega = 2\pi f = 2(3.14)(1000) = 6280$ radians/second

Substituting and differentiating v(t) we obtain:

$$i(t) = (100\,\mu F)\frac{d}{dt}(100\sin 6280t) = 100\,\mu(6280)(100)\cos \omega t = 62.8\cos(6280t) \text{ amps}$$

This result is shown in the trace of I(C1) in the preceding **PROBE** plot.

Sinusoidal Voltage Applied to an Inductor

Figure 5.02 shows a circuit that has a sinusoidal voltage source **V1** applied across an inductor. We shall perform a 2 millisecond **transient analysis** to obtain the current I(L1) through that inductor.

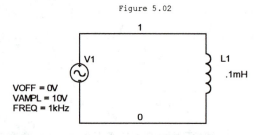

Figure 5.02

When we attempted to run the analysis, the **Output File** produced the following message:

ERROR—Voltage source and /or inductor loop involving V_V1
You may break the loop by adding a series resistance.

This message occurred because **PSpice** models an inductor like a voltage source. The circuit in Figure 5.02 has two voltage sources of unequal voltages in parallel. We shall fix the problem by adding the resistor **Rloop** with a resistance of 0.001 Ω, as shown. The addition of this small resistor will not fundamentally alter the operation of this circuit. Shown in Figure 5.03 is the modified circuit of Figure 5.02.

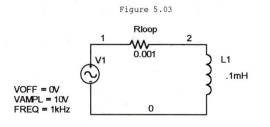

Figure 5.03

The **PROBE** plot shows the result of the analysis.

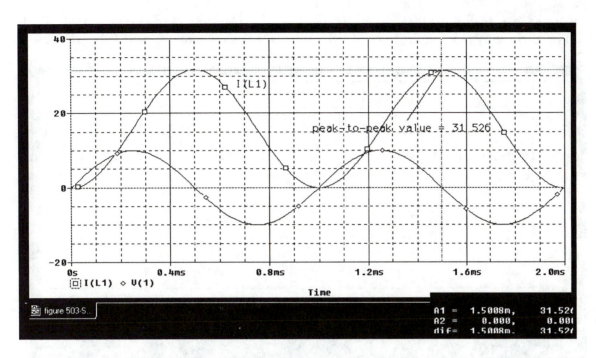

The current has a peak-to-peak amplitude of 31.52 amps and a peak amplitude of 15.92 amps. The peak amplitude is equal to the value of the arbitrary constant **K** of integration as will be shown.

For the inductor: $i(t) = \dfrac{1}{L} \int v(t)dt + \mathrm{K}$ where $v(t) = 100 \sin \omega t$

K is the arbitrary constant of integration.

Since $\omega = 2\pi f = 2(3.14)(1000) = 6280$ radians/second,

$$i(t) = \frac{1}{.1m} \int 10 \sin(6280)t \, dt + K = \frac{-1}{(6280)(.1m)} [10\cos(6280t)] = -15.92\cos(6280)t + K$$

We next evaluate K.

At t = 0 s, there was no initial current flowing in the inductor.

Therefore   i(0) = 0 = -15.92cos(0) + K

Since cos(0) = 1, therefore: 0 = -15.92 + K, and K = 15.92 amps.

Due to the offset of 15.92 amps of I(L1), it is not directly possible to determine the phase shift between V(1) and I(L1) from the above **PROBE** plot. But, with a little ingenuity, we shall nevertheless get that phase shift. The trick is to eliminate that offset. This was done by plotting the trace of [I(L1) - 15.92] as shown on the next **PROBE** plot.

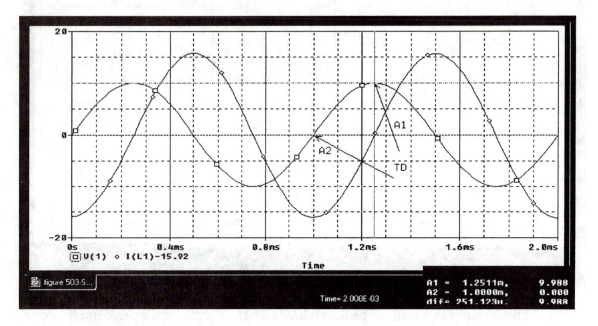

The positions of cursors **A1** and **A2** measure a **TD** of 251.123 μs between them. This translates into a 90° phase lead of the voltage V(1) relative to the inductor current I(L1).

A SINUSOIDAL VOLTAGE APPLIED TO AN RC CIRCUIT

A sinusoidal voltage with an amplitude of 100 volts and a frequency of 1kHz is applied to the circuit in Figure 5.04. Our objective is to obtain the circuit current and the circuit voltages. Of particular interest is the phase angle between the current and the voltage V(1). That phase angle is referred to as the **impedance angle** of the circuit. A **transient analysis** of 2 milliseconds will yield two cycles of the variables.

Figure 5.04

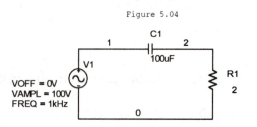

The **PROBE** plot below shows the traces of the nodal voltage V(1), which is identical to the source voltage **V1** and the circuit current. The traces of I(C1) and of I(R1) are identical to each other and to the circuit current I(V1) because in this series circuit there is only one unique current.

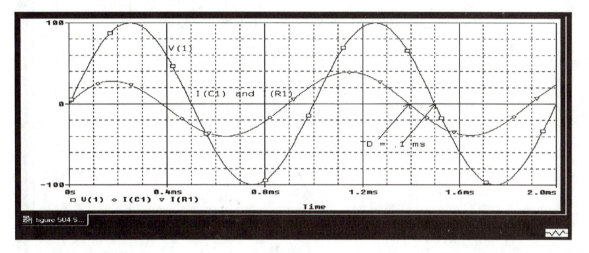

The current trace of I(C1) leads the trace of V(1) by .100 milliseconds. This converts into a 38° phase lead. We next determine the amplitude and the phase shift between the current I(C1) and the voltages V(1,2) and V(2). These two voltages are the voltages across the capacitor and the resistor respectively. The results are shown next.

The peak voltage of the capacitor voltage V(1,2) is 62 V and the peak voltage V(2) of the resistor is 78 V. The **TD** between I(C1) and V(2) is zero seconds, whereas the **TD** between I(C1) and V(1,2) is 251.155 microseconds. These **TD**s compute to a zero phase shift between the current I(C1) and the resistor voltage V(2) whereas the capacitor voltage V(1,2) lags the current by about 90°.

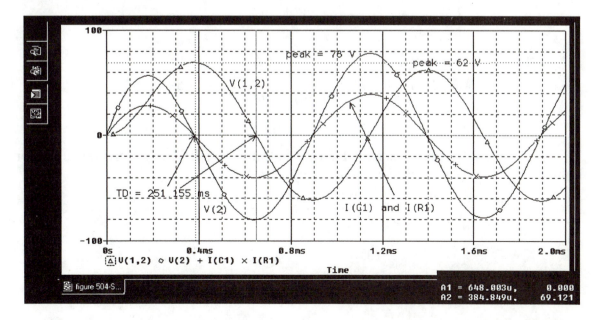

Finally, we shall determine the phase shifts between the source voltage **V1** and the voltages V(1,2) and V(2). The results are shown next. V(2) leads V(1**)** by 1.05 milliseconds. This computes to a 38° phase lead. Voltage V(1,2) lags V(1) by 1.48 milliseconds. This computes to a 52° phase lag.

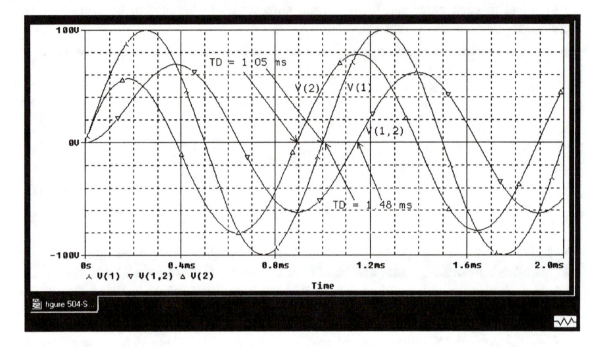

Phasor Analysis of Figure 5.04

A note on notation: Phasor currents and voltages are in **boldface** capital letters. A phasor has the first letter of its unit in a capital letter. Phasor amplitudes are equal to the **RMS** value of their associated time functions. Time functions of currents and voltages are in capital letters. The unit of a time function is spelled out in lower case letters for both source and nodal quantities.

For the input impedance of Figure 5.04 we obtain:

$$\mathbf{Z} = R - jX_c = 2 - j\frac{1}{\omega C} = 2 - j\frac{1}{(6280)(100\mu F)} = 2 - j1.59 = 2.55\angle - 38.5° \ \Omega$$

To obtain the phasor current **I(C1)** or its equivalent **I(R1),** we divide the phasor voltage of **V1**, or its nodal equivalent **V(1),** by **Z.**

Since phasor $V1 = 70\angle 0°$ V,

$$I(C1) = \frac{V1}{Z} = \frac{70\angle 0°}{2.55\angle -38.5°} = 27.5\angle 38.5° \text{ A}$$

Therefore, $V(1,2) = 27.5\angle 38.5° * 1.59\angle -90° = 43.7\angle -51.5°$ V

and $V(2) = 27.5\angle 38.5° * 2\angle 0° = 54.9\angle 38.5°$ V

These phasors and their associated angles are plotted in the phasor diagram shown.

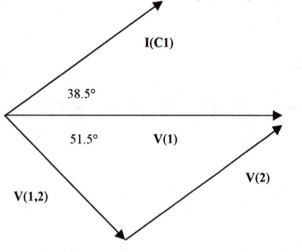

The capacitor phasor $V(1,2)$ lags phasor $I(C1)$ by 90°. Thus, it is drawn at that angle with respect to $I(C1)$. The resistor phasor $V(2)$ is in phase with $I(C1)$ and is drawn parallel with phasor $I(C1)$.

Concluding our analysis, we convert these phasors into their time functions:

$$I(R1) = I(C1) = 38.1\sin(\omega t + 38.5°) \text{ amps}$$

$$V(1) = 100\sin\omega t \text{ volts}$$

$$V(1,2) = 61.6\sin(\omega t - 51.5°) \text{ volts}$$

$$V(2) = 77.4\sin(\omega t + 38.5°) \text{ volts}$$

These results are confirmed by their **PROBE** traces.

A SINUSOIDAL VOLTAGE APPLIED TO AN RL CIRCUIT

A sinusoidal voltage source **V1** with an amplitude of 100 volts and a frequency of 1 kHz is applied to the circuit in Figure 5.05. Our objective, as it was for the circuit in Figure 5.04, is to obtain the circuit current and the circuit voltages. What is the phase angle between the circuit current and the voltage V(1)? A **transient analysis** of 2 milliseconds will yield two cycles of the variables.

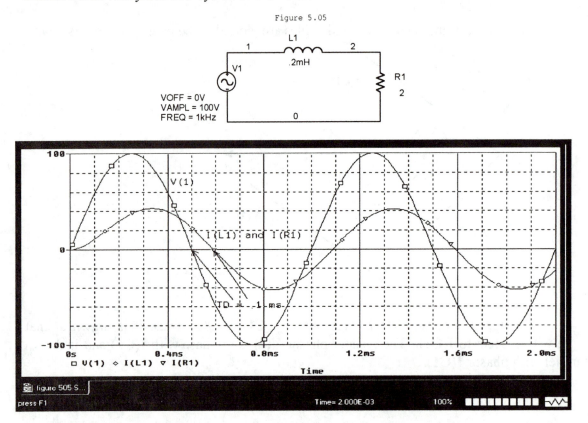

The **PROBE** plot shows that I(L1) has a **TD** of .08 milliseconds relative to V(1). This computes to a 32° phase lag. Again, notice the equivalency of the traces of I(L1) and I(R1). They are identical because there is only one unique current in this series circuit.

The next **PROBE** plot shows that the voltages V(1,2) and V(2) have voltage peaks of 53 volts and 82 volts. The **TD** between the current I(L1) and V(2) is zero seconds. This corresponds to a zero degree phase shift. The **TD** between I(L1) and V(1,2) has the latter leading the former by 249.98 microseconds. This computes to a 90° phase lead.

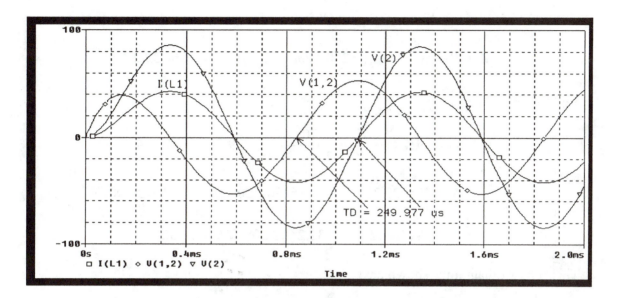

Lastly we shall determine the phase shifts between the voltage V(1), the voltage V(1,2) across the inductor and the voltage V(2) across the resistor.

The **PROBE** plot shows that voltage V(1,2) leads V(1) by a **TD** of 161.288 microseconds. This computes to a 58° phase lead. The voltage V(2) lags V(1) by a **TD** of 86.149 microseconds. This computes to a 32° phase lag.

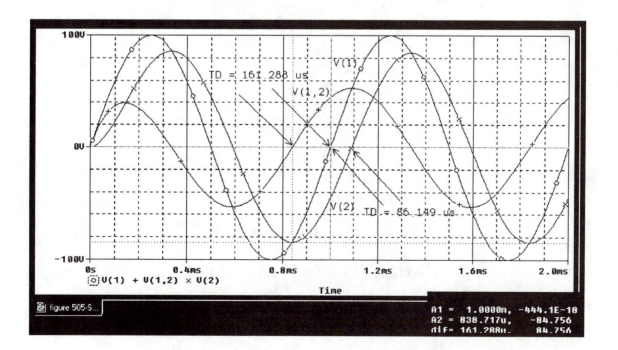

Phasor Analysis of Figure 5.05

The input impedance of Figure 5.05 is:

$$\mathbf{Z} = R + jXl = 2 + j((6280)(.2m)) = 2 + j1.26 = 2.36\angle32.2°\ \Omega$$

Computing the phasor current:

$$\mathbf{I(L1)} = \frac{V1}{Z} = \frac{70\angle0°}{2.36\angle32.2°} = 29.7\angle-32.2°\ \text{A}$$

and:

$$\mathbf{V(1,2)} = 29.7\angle-32.2° * (6280)(.2m)\angle90° = 37.3\angle57.8°\ \text{V}$$

$$\mathbf{V(2)} = 29.7\angle-32.2° * 2\angle0° = 59.4\angle-32.2°\ \ \text{V}$$

Their phasor diagram is shown next:

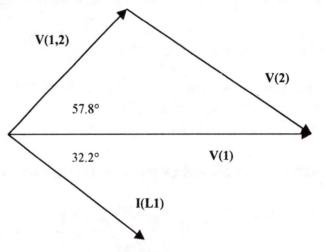

We convert the phasors to their corresponding time functions:

$$V(1) = 100\ \sin\omega t\ \text{volts}$$

$$I(L1) = I(R1) = 41.8\ \sin(\omega t - 32.2°)\ \text{amps}$$

$$V(1,2) = 52.5\ \sin(\omega t + 57.8°)\ \text{volts}$$

$$V(2) = 83.7\ \sin(\omega t - 32.2°)\ \text{volts}$$

These results are consistent with the **PROBE** data.

A SINUSOIDAL VOLTAGE APPLIED TO A SERIES RLC CIRCUIT

The circuit in Figure 5.06 contains a resistor, a capacitor and an inductor. The circuit can store both electrical and magnetic energy. These are the only forms of energy that can be stored in any electrical circuit. Thus, this circuit assumes great importance in our study. This is especially true if we consider that the circuit in Figure 5.06 can be representative of many, far more complex **RLC** circuits. Lastly, the ability to store two independent forms of energy gives rise to the phenomena of resonance. Its importance can not be overstated and shall occupy us for a considerable amount of time in subsequent chapters.

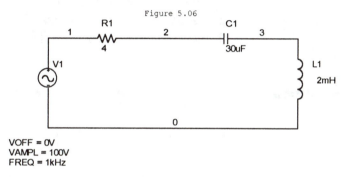

Figure 5.06

VOFF = 0V
VAMPL = 100V
FREQ = 1kHz

A **transient analysis** of 2 milliseconds produced the **PROBE** plot of the source voltage **V1** or its nodal equivalent V(1) and the circuit current I(R1). That current is also equal to I(C1) and I(L1).

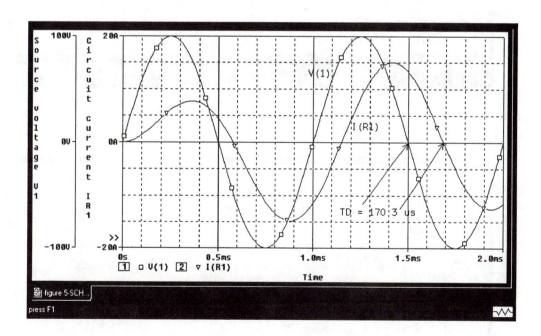

V(1) and the circuit current I(R1) have been referenced to two **Y-axes** for reason of scale. The **TD** has been determined after V(1) and I(R1) assumed their steady-state values. It was measured by the two cursors **A1** and **A2.** Its value is 170.3 microseconds. This calculates to a 61.3° phase lead of V(1) relative to I(R1).

The next **PROBE** plot shows the traces of the circuit voltages. The inductor voltage V(3) leads the resistor voltage V(1,2) by 90° and the capacitor voltage V(2,3) lags V(1,2) by 90°. This makes the phase shift between V(3) and V(2,3) 180°. Whenever V(3) has a positive peak, V(2,3) has a negative one and vice versa. These two voltages are said to be out of phase. This is a most important characteristic and will lead to the phenomenon of resonance.

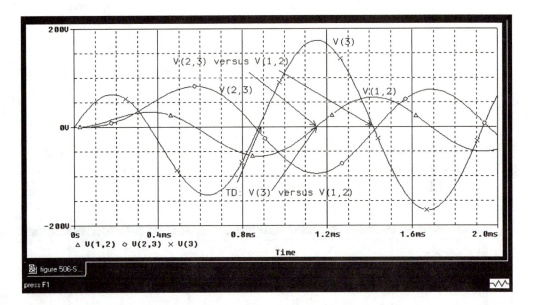

Let us plot the source voltage V(1), the inductor voltage V(3) and the ratio RMS(V(3))/RMS(V(1)).

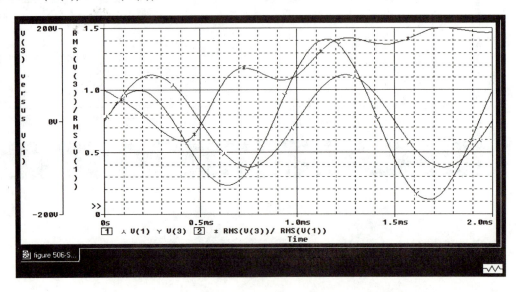

It shows that this ratio approaches 1.45 by about 2 milliseconds. Thus, a circuit voltage is larger than the source voltage! This points to the possibility of voltage amplification. We shall return to this fact when we investigate the frequency response of this circuit. For now, the possible voltage levels encountered in such a circuit calls for caution on the part of operating personnel in a laboratory setting.

Phasor Analysis of Figure 5.06

$$\mathbf{Z} = R1 + j(\omega L - \frac{1}{\omega C}) = 4 + j(12.6 - 5.3) = 8.3\angle 61.3° \ \ \Omega$$

$$\mathbf{I(R1)} = \frac{V(1)}{Z} = \frac{70\angle 0°}{8.3\angle 61.3°} = 8.4\angle - 61.3° \ \ A$$

$$\mathbf{V(1,2) = I(R1)*R(1)} = 8.4\angle - 61.3° * 4\angle 0° = 33.6\angle - 61.3° \ \ V$$

$$\mathbf{V(2,3) = I(R1)*(-jXc)} = 8.4\angle -61.3° * 5.3\angle -90° = 44.6\angle -151.3° \ V$$

$$\mathbf{V(3) = I(R1)*(jXl)} = 8.4\angle -61.3° * 12.6\angle 90° = 106\angle 28.7° \ V$$

$$\mathbf{V(1)} = 70\angle 0° \ V$$

Their phasor diagram is shown

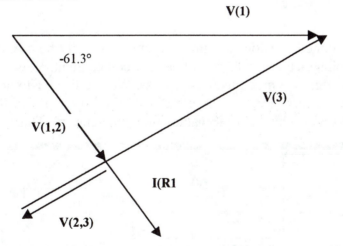

Converting the phasors to time functions, we obtain:

$$I(R1) = 11.7 \sin(\omega t - 61.3°) \text{ amps}$$

$$V(1,2) = 47.4 \sin(\omega t - 61.3°) \text{ volts}$$

$$V(2,3) = 62.9 \sin(\omega t - 151.6°) \text{ volts}$$

$$V(3) = 149.5 \sin(\omega t + 28.7°) \text{ volts}$$

Determining the Impedances of Figure 5.06 Using PROBE

In the phasor analysis of this circuit, we calculated the impedances of **Z**, **R1**, **L1** and **C1** from their formulations given. The impedance of a circuit element, symbolized by **Z,** having units of ohms, is by definition the ratio of a voltage across that circuit element divided by the current though it. In the case of **R1**, that ratio was the resistance of **R1**. In the case of the determination of the capacitive and inductive reactances, the voltage/current ratios resulted in a quantity calculated from $1/\omega C$ for the former, and ωL for the latter reactance. In addition, an angle was associated with each of them. These angles are called the impedance angles of the capacitor and the inductor. Together with the reactances of the elements, they make up the impedance of the circuit elements. The reader is referred to the introductory section of this chapter.

Finally, an important parameter of a circuit is its input impedance. It is the impedance perceived by the voltage source **V1**. For Figure 5.06, it was calculated as the phasor sum of the impedances of the circuit elements. Its value, the reader will recall, was $8.3\angle61.3°\ \Omega$. The input impedance, together with the voltage source **V1** determines the total power dissipated in the circuit and the total electrical and magnetic energy stored in the circuit.

The input impedance for this simple circuit can easily be calculated, as was done above, from the voltage-current relations of each circuit element. However, with increasing circuit complexity, this becomes a tedious process. We use **PSpice** to do the job for us. To determine the input impedance, we obtain the **PROBE** trace of the **RMS** values of the source voltage **V1** divided by the **RMS** value of the total current. This we shall do next.

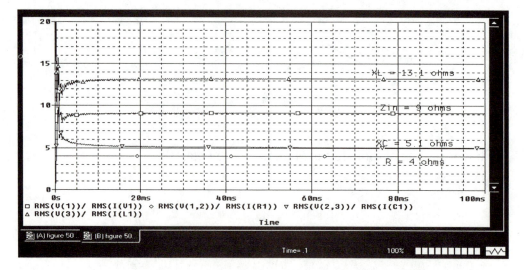

A **transient analysis** of 100 milliseconds produced the traces just shown. The reader will note that the lengthening of the **transient analysis** allowed the perturbations of the traces caused by the transients in the circuit to die out and allowed the input impedance and the reactances to attain their steady-state values. The values are reasonably close to those calculated. These values, together with their phase angles determined from the previous **PROBE** plot containing their **TDs**, are the impedances of the circuit elements in addition to the input impedance of Figure 5.06.

Adding Phase Shift to the Input Voltage V1

The circuit in Figure 5.07 is used to find out what happens to the circuit current and the circuit voltages as the source voltage **V1** is shifted by a 45° phase lag. This circuit is the same as that in Figure 5.06 except for the phase shift of **V1**.

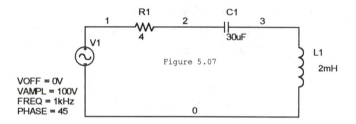

Figure 5.07

The **PROBE** plot shows that the source voltage **V1**, or its nodal equivalent V(1), and the circuit current I(R1) have been shifted to the right on the time axis by the same amount. Thus, their relative phase angle has not changed from that in Figure 5.06. It can be shown that all the voltage phasors in the circuit will have experienced a delay of 45° so that their relative phase shifts have not been changed. The reader need only to visualize that the phasor diagram shown for Figure 5.06 was rotated 45° in a clockwise direction. The proof of this is left as an exercise for the reader.

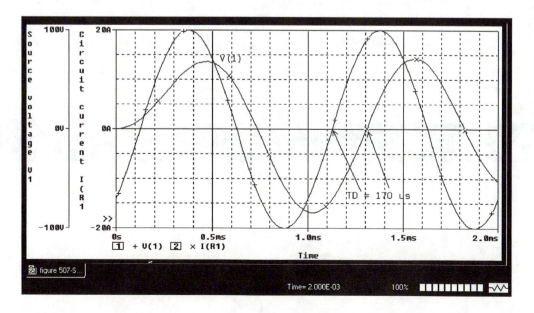

A SINUSOIDAL VOLTAGE APPLIED TO A PARALLEL RLC CIRCUIT

Figure 5.08

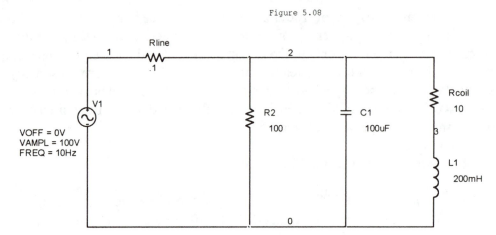

In this circuit Figure 5.08, **Rline** was added to facilitate the plotting of the total current into the circuit. Remember that I(V1) is the negative of I(Rline) due to the syntax of the **PSpice** program.   Resistor **Rcoil** has been added to represent the unavoidable resistance associated with an inductor. Also, without this resistor, the circuit will take a long time to settle into its steady-state condition. A **transient analysis** of 200 milliseconds produced the plot of the voltage V(1) and the total current I(Rline).

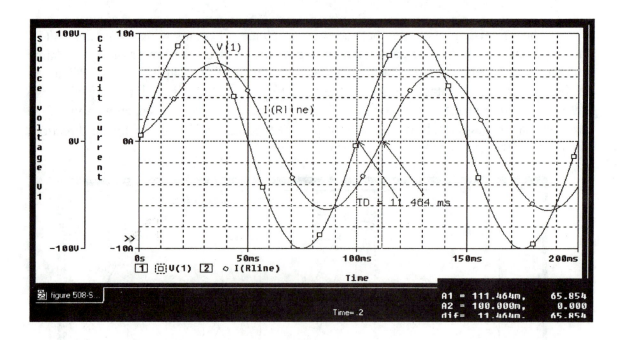

The circuit current I(Rline) lags **V1**, or its nodal equivalent V(1), by a **TD** of 11.464 milliseconds. This calculates into a 41.2° phase lag. An experienced professional would notice that the circuit is inductive, meaning that the amount of magnetic energy stored over a cycle of **V1** exceeds the electrical energy stored in the capacitor for the same time period. We shall verify this when we compare the inductive to the capacitive reactance.

The next **PROBE** plot contains the traces of the branch currents. I(R2) and I(C1) are referenced to a second **Y-axis** because their amplitudes are relatively small compared to I(L1**)**. I(R2) is in phase with V(1**),** as can be seen if we compare their traces on the two **PROBE** plots.. I(L1) lags I(Rline**)** by 2.5 milliseconds, or 9°. Therefore it lags V(1) by - 41.2° - 9° = - 50.2°. I(C1) leads I(R2) by 25 milliseconds, or 90°. Therefore, it leads V(1) by 90°. Their phasor diagram follows the **PROBE** plot shown below.

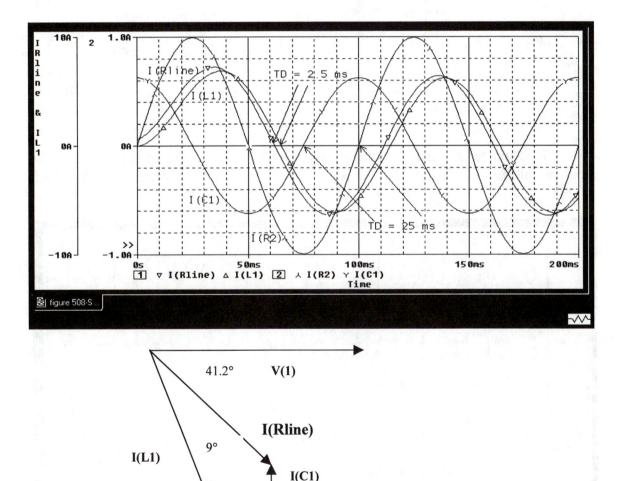

REDUCTION AND EQUIVALENCY BETWEEN ELECTRICAL CIRCUITS

What's new?

1. The use of the cosine **(COS)** and sine **(SIN)** functions.

From the perspective of a current or voltage source, there is an infinite variety of circuits that establish a particular voltage across a current source or extract a particular current from a voltage source. Such circuits all have the same input impedance or input admittance. They are electrically equivalent. It is often advantageous to find the simplest circuit that represents a more complex one. Such a substitution results in a circuit having fewer elements. This results in a simpler analysis of the circuit. We shall use the circuit of Figure 5.09 to demonstrate the design of an equivalent circuit.

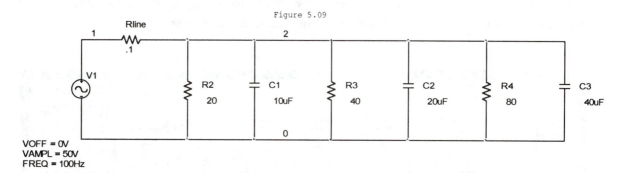

Figure 5.09

To begin our analysis , we obtain the phase shift between the voltage V(1) and the input current I(Rline). The indicated **TD** of .745 milliseconds corresponds to a phase shift of 27°.

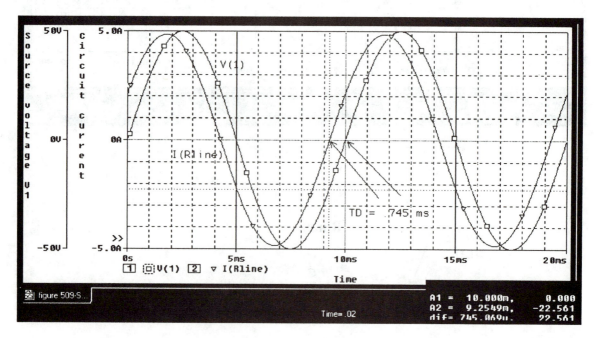

Since Figure 5.09 is a parallel circuit, we shall attempt to find an equivalent parallel circuit, although such is not necessary. We could also find a series circuit that is the equivalent of the circuit in Figure 5.09. The admittance of the equivalent circuit is expressed as:

$$\mathbf{Y} = G + jB = Y\angle\theta$$

where $G = \text{Real}(\mathbf{Y}) = Y\cos\theta$ $B = \text{Im}(\mathbf{Y}) = Y\sin\theta$ $\angle\theta = \arctan(B/G)$

In this formulation, $\mathbf{Y}$ is the admittance, G is the real part of the admittance, defined as the conductance of $\mathbf{Y}$, and B is the reactive part of the admittance, defined as the susceptance of $\mathbf{Y}$.   All quantities have the units of S (siemens). The **PROBE** plot below shows the traces of each of them.

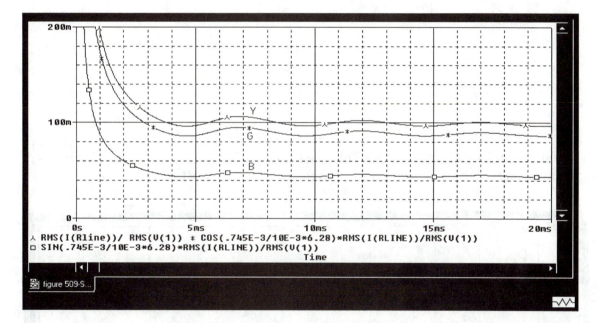

PSpice has the ability to obtain the cosine (**COS**) and sine (**SIN**) functions. They are found in the **Macros** box. You can also simply type **COS** or **SIN** to activate them.
The ratio .745E-3/10E-3*6.28 expresses the **TD** in radians. This is a must for the **COS** and **Sin** functions. The ratio RMS(I(Rline))/RMS(V(1)) is the amplitude of the admittance of the circuit. We obtain the amplitudes of Y, G and B at t = 2 milliseconds from the **PROBE** plot. At that time, the **RMS** values of the expressions have attained steady state.

We determine the equivalent circuit components from the admittance values:

$$B = \omega C = 2\pi fC = 2(3.14)(100)C = 52 \text{ mS}$$

$$\text{Therefore } C = \frac{52mS}{2(3.14)(100)} = 82.8\,\mu F$$

Since there is no circuit element that represents the conductance G, we use an equivalent resistor Req of value 1/G to represent that conductance.

$$G = 84 \text{ mS} = 1/\text{Req}; \text{ therefore, Req} = 1/84 \text{ mS} = 11.9 \ \Omega$$

The equivalent circuit is shown in Figure 5.10.

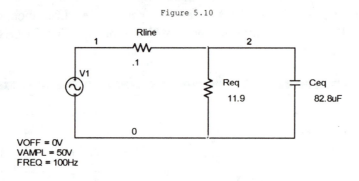

Figure 5.10

The **PROBE** plot of the **transient analysis** of this circuit is shown. The traces of V(1) and I(Rline) are identical to those obtained for the circuit in Figure 5.09. That proves the equivalency of the two circuits relative to the voltage source **V1**.

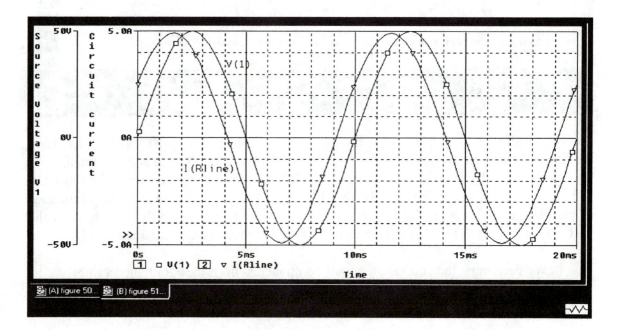

We can also obtain a series equivalent circuit of the circuit in Figure 5.09. The impedance of the equivalent circuit is expressed as:

$$\mathbf{Z} = R - jX$$

where $R = \mathrm{Re}(\mathbf{Z}) = Z \cos \theta$ $X = \mathrm{Im}(\mathbf{Z}) = Z \sin \theta$ $\angle\theta = \arctan(X/R)$

Thus, we need to find R and X. We run the **PSpice** analysis for Figure 5.09 again to obtain the traces of R and X .

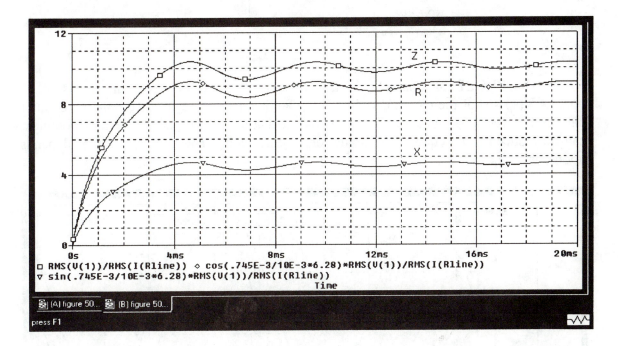

We convert the reactances to circuit elements:

$$Xc = \frac{1}{\omega C} = \frac{1}{2(3.14)(100)} = 4.8 \ \Omega$$

Solving for C

$$C = \frac{1}{2(3.14)(100)(4.8)} = 331 \mu F$$

from the **PROBE** plot

$$R = 8.2 \ \Omega$$

The equivalent circuit is shown in Figure 5.11.

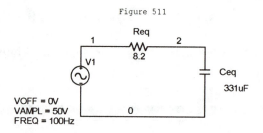

The **PROBE** plot based on this equivalent circuit shows that the traces of its voltage V(1) and of the circuit current I (Req) are identical to their corresponding counter parts in Figures 5.09 and 5.10. Thus we observe that any circuit, no matter how complex, can be reduced to an equivalent parallel or series circuit. The choice of either depends upon one's objectives.

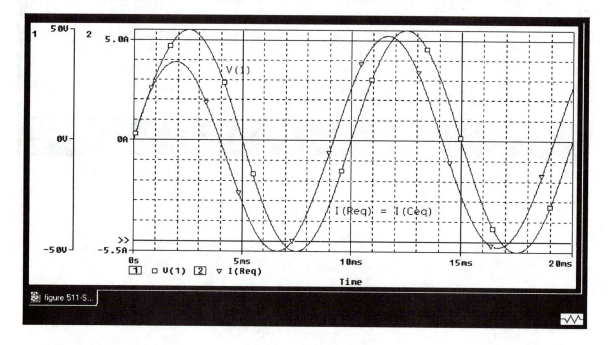

PROBLEMS

5.1   For the circuit shown:
       (a). Find all nodal voltages
       (b). Find the time delays **(TD)** and the corresponding phase shifts between the
             voltages across **R1** and **C1** relative to the source voltage **V1**.
       (c). Obtain the **RMS** values of all voltages.
       (d). Do a phasor analysis of the circuit voltages.
       (e). Draw a phasor diagram of these voltages.
       (f). Obtain the time functions of the circuit voltages.
       (g). Compare the time functions obtained from the phasor analysis with those
             obtained from their **PROBE** traces.

Problem 5.1

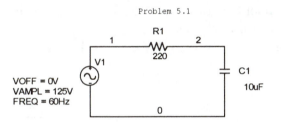

5.2   The voltage source **V1** has a 4 V offset value, a peak value of 4 V and a
       frequency of 1 kHz.
       (a). Find the voltages V(R1) and V(C1).
       (b). Find the circuit current I(R1).
       (c). Prove that I(R1) = I(C1) = minus I(V1).
       (d). Determine the **TD's** of V(R1**)** and V(C1) relative to V(1).
       (e). Calculate the phase shifts of V(R1) and V(C1) relative to V(1).
       (f). What, if any, is the average voltage V(1,2), Hint:Use the **AVG** function?

Problem 5.2

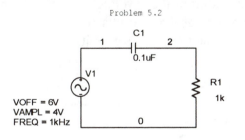

5.3   Repeat Problem 5.2 with the positions of the capacitor and the resistor interchanged. Compare the results of the two problems and discuss the reason(s) for the different results.

5.4   This circuit was constructed by a somewhat inexperienced technician, so unlike you. When a voltmeter was placed between nodes 3 and 0, it became obvious that something was wrong. Perform a **PSpice** analysis to find out what the voltage across those two nodes is.

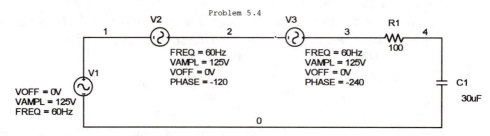

Problem 5.4

5.5   Find the amplitude of the voltage V(1,2) across the current source **I1**. Also, find the phase angle between that voltage and the current of source **I1**. Remember: every node in **PSpice** must have a dc path to ground (node 0).

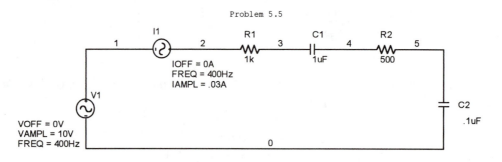

Problem 5.5

5.6   For the circuit shown:
(a). Find the amplitudes of the voltage across **R1** and **L1**.
(b). Find the phase shift of these voltages relative to the voltage **V1.**
(c). Perform an phasor analysis of this circuit.
(d). Draw a phasor diagram of this circuit.

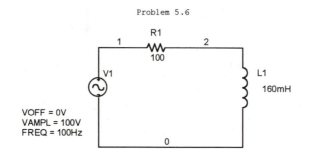

Problem 5.6

5.7 For the circuit, determine the amplitude and the phase angle of the input impedance. Find an equivalent circuit that consists of one resistor and one inductor.

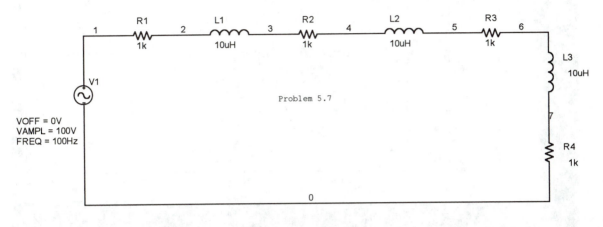

Problem 5.7

5.8 Select an inductance for this circuit so that its input impedance angle is 30°. Check your design with a PSpice analysis.

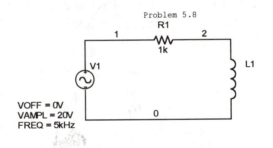

Problem 5.8

5.9 This circuit was to have a voltage V(2) equal to 70% of the voltage of V(1). Also, the circuit was to provide a 45° phase lead of V(2) relative to V(1) at the applied frequency of 600 Hz. When a **PSpice** analysis was run, the **PROBE** plot of the traces of V(1) and V(2) was produced. Clearly, neither of the two criteria has been satisfied. A **TD** of .390 milliseconds corresponds to an 84.5° phase shift. An ohmmeter showed that the value of the resistor was correct. Replace the inductor with one that has the correct inductance for this circuit to perform as specified.

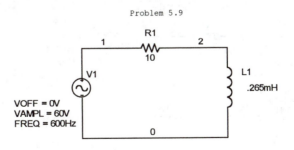

Problem 5.9

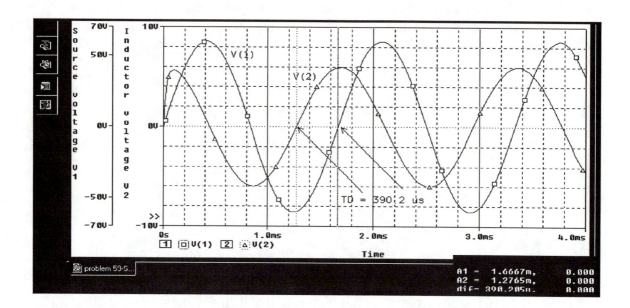

5.10   For this circuit, determine the values of **R1** and **L1** so that the voltage V(2)
        leads voltage V(1) by 30° and the ratio of RMS(V(1))/RMS(I(R1)) = 100.
        Check your design with a **PSpice** analysis.

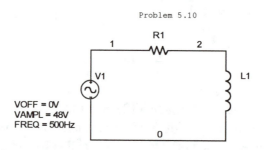

Problem 5.10

VOFF = 0V
VAMPL = 48V
FREQ = 500Hz

5.11   The circuit shown has two voltage sources, **V1** and **V2**. Both are operating
        at 60 Hz. **V2** lags **V1** by 90°. Find the current I(R1) and the voltages
        V(1,2) and V(2,3). What are their phase angles relative to V(1)?

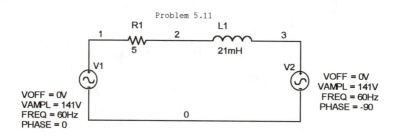

Problem 5.11

5.12   The voltage source **V2** now has phase lead of 90°. All else remains the
same as in Problem 5.11. Find the current I(R1) and the voltages V(1,2)
and V(2,3). What are their phase angles relative to V(1)? Compare the
results of the two problems in regard to their respective phase shifts.

5.13   For this circuit, find the voltages across **R1**, **L1** and **C1**. Find the impedance
angle of these voltages. In particular, from the angle of the input impedance
determine which of the two reactances is the larger one.

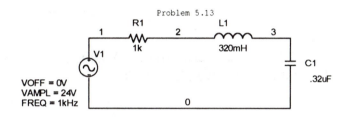

5.14   Obtain the input impedance, both its amplitude and phase angle, of the
RLC network shown. The amplitude of **V1** is left for the reader to decide.

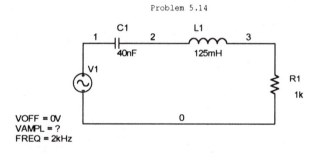

5.15   The frequency of voltage source **V1** in Problem 5.14 has been changed to 4kHz.
For this new parameter, find the   input impedance and compare it to that in
Problem 5.14. In particular, which reactance got smaller and which got bigger?
Explain the values of the two reactances for the conditions when the frequency
approaches 0 Hz and infinity Hz.

5.16   For this circuit:
       (a). Find the amplitude and phase angles of all circuit voltages.
       (b). Find the amplitude and the phase angle of the current.
       (c). What is the sum of the inductor and the capacitor voltages?
       (d). Obtain the input impedance of the circuit.
       (e). What is the operating condition of this circuit called?
       (f). What is the total power into this circuit?

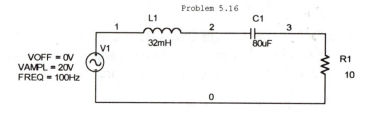

5.17   The given **PROBE** data are from a parallel **RL** circuit. From that data, synthesize
       the simplest circuit that will give the same data.

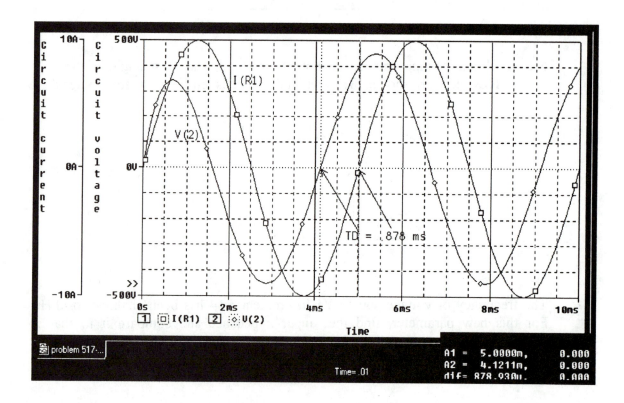

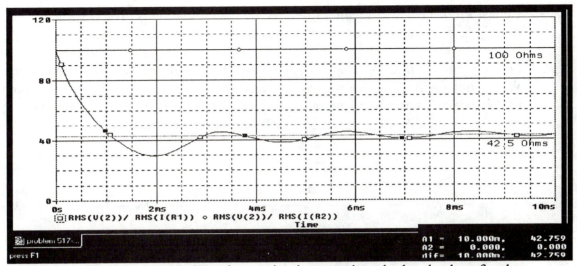

At the completion of the design of your circuit, enter the calculated values for the
R(2) and **L(1)** into the circuit shown. Run a **PSpice** analysis to check if you get the
same data as above.

Problem 5.17

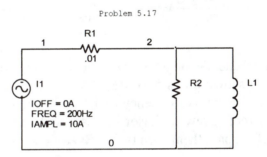

5.18 For this circuit:
  (a). Find all circuit voltages and the current.
  (b). Find all **TD**s and the corresponding phase shifts.
  (c). Determine the amplitude and the phase angle of the input impedance.
  (d). Is the circuit capacitive or inductive?
  (e). What is the total power consumption per cycle of the input voltage?
  (f). Obtain a phasor diagram of this circuit.
  (g). From the phasor diagram, obtain the time functions of the circuit voltages and
      the current.

Problem 5.18

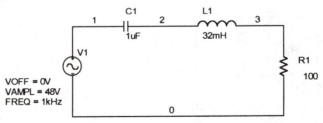

5.19   Repeat Problem 5.18 but with V1 delayed by .42 milliseconds. Answer all
        questions as in that problem. Compare the answers of the two problems.

5.20   For this circuit, find the amplitude and the phase angle of the input impedance.
        Watch out for floating nodes! Note: Since the requirement is to find the input
        impedance and not any of the individual voltages across the various elements, it
        may pay to simplify this circuit before a **PSpice** analysis is run.

Problem 5.20

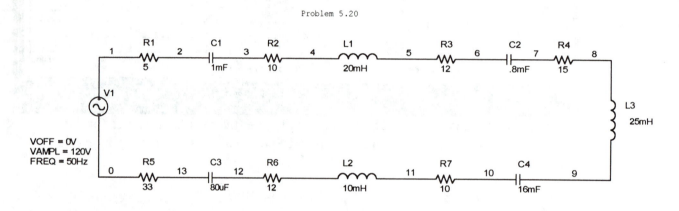

5.21   **V1** has a frequency of .001 Hz. The resulting voltage V(3) across the
        capacitor is shown on its **PROBE** plot. Find the amplitude of the voltage
        source **V1**. Change its frequency to 1000 Hz and obtain all circuit voltages
        and the current. Draw a phasor diagram and convert the phasors to time
        functions. Compare them with their **PROBE** traces.

Problem 5.21

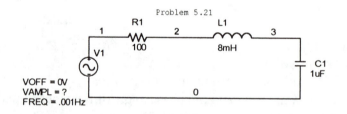

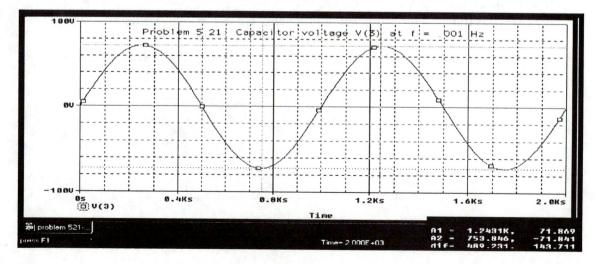

5.22 For this circuit, the **PROBE** plot contains the traces of the voltages V(1), V(1,2) and V(3). From this data determine the value of the capacitor **C1.** Check the correctness of your analysis with a **PSpice** run. Hint: This circuit is at resonance.

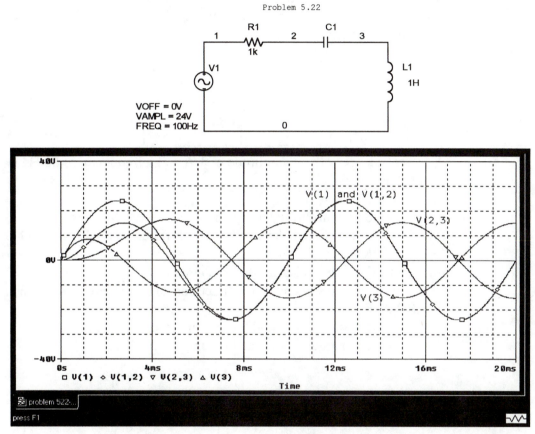

Problem 5.22

5.23 This **RC** circuit has a voltage source of 48 volts amplitude and a frequency of 100 Hz applied to it. Find the following:
(a). The current into the circuit and through **R1** and **C1**
(b). The **TD** (time delay) and phase shift between V(1) and I(V1**)**
(c). The admittance and impedance of the circuit
(d). The phasor diagram
(e). The time functions derived from the phasor diagram

Problem 5.23

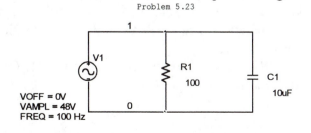

5.24   A current source of amplitude 100 mA and a frequency of 1000 Hz is
applied to this circuit. Find the following:
(a) The total current into the circuit and the currents through **R1** and **L1**
(b) The **TD**(time delay) and phase shift between V(1) and I(V1)
(c) The admittance and impedance of the circuit
(d) The phasor diagram and corresponding time functions
(e) Why was there no loop error in this circuit?

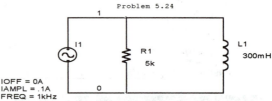

5.25   A current source with an amplitude of 1 A and a frequency of 60 Hz is
applied and shown.
(a). Find the voltage V(1).
(b). Find the current through each circuit element.
(c).   Find the **TD** (time delay) and phase shift between I(I1) and V(1).
(d). Verify Kirchhoff's current law.
(e).   Draw the phasor diagram for this circuit.
(f). Convert the phasors into time functions and compare them to their
PROBE traces.
(g). Find the conductances and the susceptances.
(h). Find the input admittance and the input impedance.

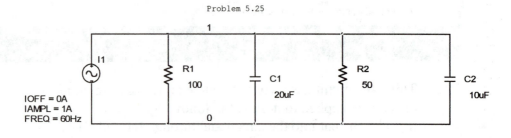

5.26   This circuit resembles a linear small-signal BJT (Bi-polar transistor)
common emitter amplifier circuit. It contains a current controlled current
source **F1**. The current gain is equal to 100.
(a). Find the voltages V(1) and V(4).
(b). Obtain the ratio RMS(V(4))/RMS(V(1)). That ratio is the voltage gain
of the amplifier.
(c). What is the phase shift between V(1**)** and V(4)? Can you explain it?
(d). Find the current through **Rin.**

(e). Find the sum of the currents through **RC** and **Rload.**

(f). Find the ratio of RMS(I(RC + Rload))/RMS(Rin)). It is defined as the current gain of the amplifier.

(g). Does it agree with the indicated current gain of the source **F1**?

(h). What are the effects of **Cin** and **Cout** at this frequency? Could we have removed them without causing any major change in the circuit operation?

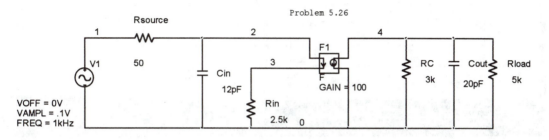

Problem 5.26

5.27 Given the **PROBE** data shown for this circuit, find the values of **R1** and **C1.**

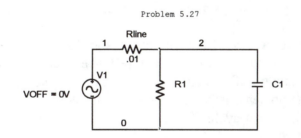

Problem 5.27

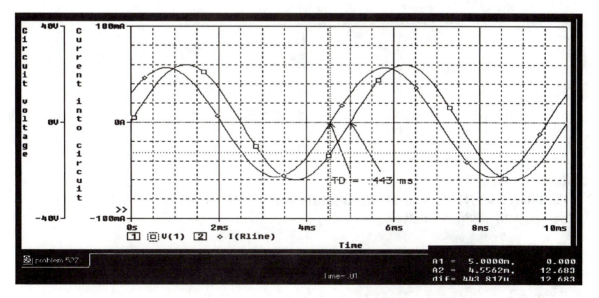

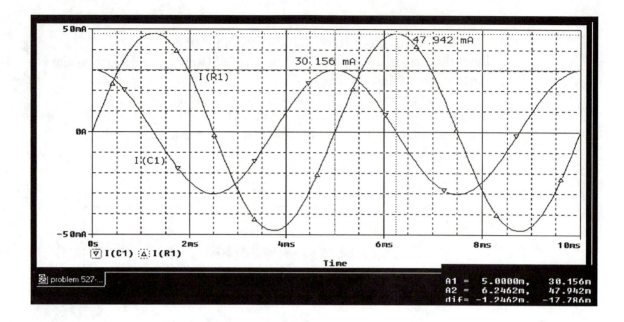

From the **PROBE** data, enter the voltage of **V1**, its frequency and the calculated values of **R1** and **C1** in the circuit shown. **Rline** was added to obtain a trace of the total current **I(Rline)** into the circuit without an extraneous phase shift relative to the voltage **V1**. Run a **PSpice analysis** to confirm your calculations.

5.28   For this circuit, find the voltage V(1) and the instantaneous current through each inductor. Find the input admittance and input impedance. Determine their phase angle. Plot the phasor diagram of this circuit and obtain the time functions from that diagram. What is the power dissipated in this circuit? How much of magnetic energy is stored in this circuit per cycle of **I1?**

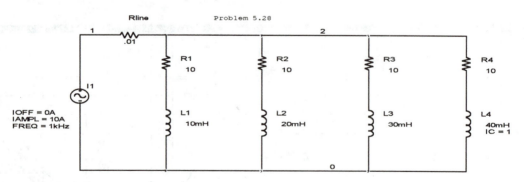

5.29   Find the series **RL** equivalent circuit for the parallel **RL** circuit shown. Check your design with a **PSpice** analysis.

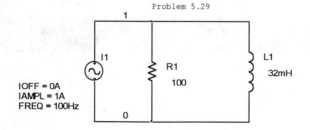

5.30   The circuit has two sinusoidal current sources, **I1** and **I2**. Both have a frequency of 10 kHz. Find the circuit voltage and the currents through each branch of the circuit. Verify Kirchhoff's current law at node 2. Rline 1 and Rline 2 have been added to ease the determination of the source currents into the circuit.

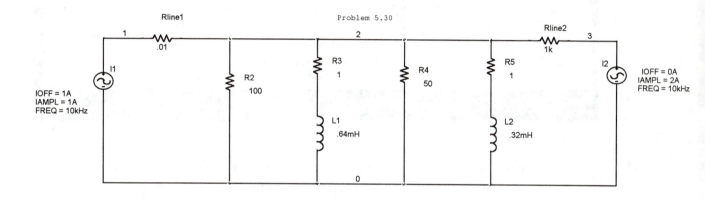

5.31   In Problem 5.30, the current source **I2** is delayed by 25 microseconds. Find all currents in the circuit and verify Kirchhoff's current law at node 2. Compare your present results with those obtained in Problem 5.30.

5.32   From the given **PROBE** data, synthesize the simplest parallel **RL** circuit. Check your design with a **PSpice** analysis. Use a sinusoidal current source to drive the circuit.

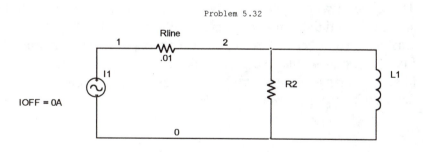

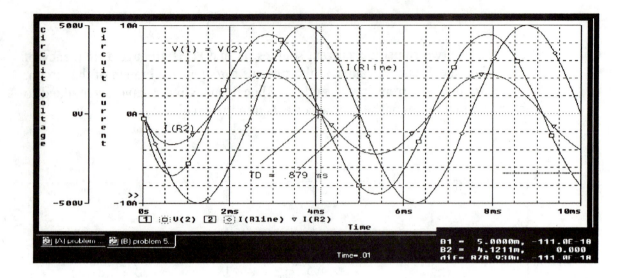

5.33   Find the input admittance and the input impedance of this circuit. Also, from your results, what is the equivalent inductance of two inductors in parallel? What the equivalent resistance of two resistors in parallel?

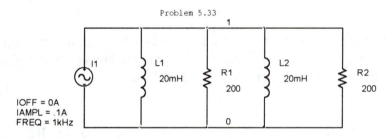

Problem 5.33

5.34   This circuit was designed to provide a 45° phase shift between the voltage source **V1** and the current I(Rline). When the **PROBE** plot shown below was obtained, it became evident that the circuit did not perform as intended. Fix the design so that the proper phase shift will be obtained. You are at liberty to change either or both of the passive circuit elements. Check your design with a **PSpice** run.

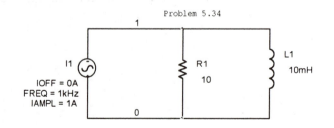

Problem 5.34

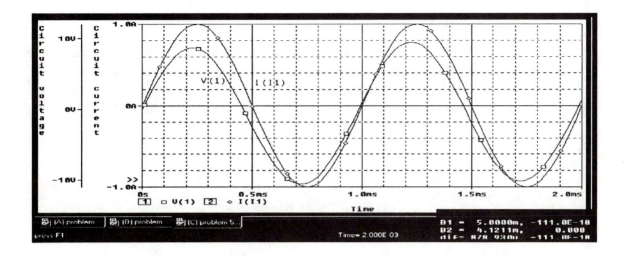

5.35 For this circuit, find the simplest equivalent series circuit. Check your
design with a **PSpice** analysis.

Problem 5.35

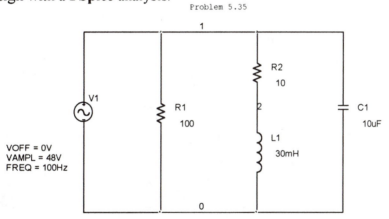

5.36 For this circuit, find the simplest equivalent series circuit. Check your
design with a **PSpice** analysis.

Problem 5.36

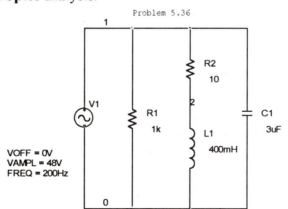

5.37   Repeat Problem 5.36 at a frequency of 400 Hz. Compare the two equivalent
        circuits. The important point to note is that the two equivalent circuits
        differ because of the change in frequency. This dependency of circuits on
        frequency will receive extensive coverage in future chapters.

TOTAL RESPONSE OF RC, RL AND RLC CIRCUITS WITH SINUSOIDAL SOURCES

INTRODUCTION

In the last chapter, we obtained the **steady-state response** of series and parallel **RC, RL** and **RLC** circuits to sinusoidal current and voltage sources. In each case analyzed, we observed that any of the resultant circuit currents and voltages resembled the source current or voltage, differing at most in amplitude and phase shift. The term **steady-state, or forced, response** refers to that condition.

However, this is not the total response of a network when a current or voltage source is applied to it. From the moment at which such a source is applied to a network, it takes a finite time until all the circuit variables resemble the source function. That time and the manner in which the variables change to conform to the source function is defined as the **transient, or natural, response** of the network. The **natural response** is dependent upon the elements in a circuit, their initial energy content or lack of it and the manner in which these elements are connected.

The Total Response of an RC Circuit

The **total response** of a circuit is defined as the sum of the **transient response** plus the **steady-state response** of the circuit. The **PSpice** program calculates them both and the **PROBE** plots displays the results graphically. Real-world circuits are often designed to make the **transient response** die out quickly. This was also done for the circuits in the previous chapter in which the transients lasted no more than a fraction of the period of a sinusoidal current or voltage. Presently, the values of circuit elements are chosen to make the **transient response** apparent. We shall obtain both the **transient and the steady-state responses** of the circuit in Figure 6.01.

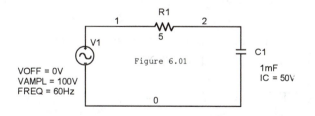

The voltage source **V1** is applied to the circuit at t = 0 seconds. The initial voltage across the capacitor **C1** is 50 volts. We shall obtain the circuit voltages and current by running a 60 milliseconds **transient analysis**. The analysis produced the traces of V(1)

and V(2) shown on the **PROBE** plot.

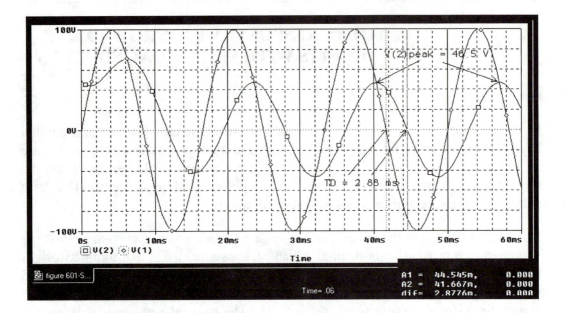

The capacitor voltage V(2) starts from its initial value of 50 volts and decreases to its **steady-state** voltage in about 25 milliseconds. Thereafter, its peak voltage remains at 46.5 volts. The reader will recall that the product RC determines the time constant of a RC circuit. For this circuit, this calculates to 5 milliseconds. We also recall that **steady-state conditions** were attained in about five time constants. This, for our circuit, is equal to 25 milliseconds. The **PROBE** plot confirms this. The source voltage **V1**, or its nodal equivalent V(1), leads the capacitor voltage V(2) by 2.88 milliseconds. This calculates to a phase lead of 62.2°.

Let us perform an analysis of this circuit "on paper" to obtain V(2). **The total response** is the **forced response** due to the voltage source V(1) and **the transient, or natural, response**. This is expressed symbolically as follows:

$$V(2) = V(2)_F + V(2)_N \text{ volts}$$

The subscripts F and N refer to the **forced** and the **natural responses** respectively. We obtain the **forced response** using phasor analysis.

$$V(2)_F = \frac{(-jX_C)(V(1))}{(R - jX_C)} = \frac{(0 - j2.65)(70 - j0)}{(5 - j2.65)} = 15.5 - j29 = 32.8\angle -62.2° \text{ V}$$

Its associated time function is $V(2)_F = 46.5 \sin(\omega t - 62.2°)$ volts.

To obtain the **transient, or natural, response**, we need to set V(1) to zero volts. Applying Kirchhoff's voltage law:

$$V(1,2)_N + V(2)_N = 0 \text{ volts} \qquad \text{hence} \quad R(1)*I(R1)_N + V(2)_N = 0 \text{ volts}$$

$$\text{but} \quad I(R1) = C\frac{d}{dt}(V(2)) \text{ amps}$$

$$\text{therefore} \quad RC\frac{d}{dt}(V(2)) + V(2) = 0 \text{ volts}, \quad t > 0 \text{ seconds}$$

This differential equation governs the **natural response** of the capacitor voltage $V(2)_N$ for any time after zero seconds. The second derivative of $V(2)_N$ is proportional to $V(2)_N$. This allows us to assume an exponential equation format for $V(2)_N$.

$$V(2)_N = Ae^{st} \text{ volts}$$

It remains to determine the constant s and A. We substitute our assumed solution into our differential equation and obtain:

$$RC(sAe^{st}) + Ae^{st} = 0 \text{ volts}, \qquad t > 0 \text{ seconds}$$

We simplify this equation by factoring the common term and divide both sides of the equation by **RC**.

$$(s + 200)Ae^{st} = 0 \text{ volts}$$

For the product of two factors to be equal to zero, at least one of them must be equal to zero. The nontrivial solution is that the quantity enclosed in parentheses is equal to zero. Thus,

$$s + 200 = 0$$

This equation is defined as the **characteristic equation** of our circuit. Its solution yields the value of s.

$$s = -200$$

We enter this value into our assumed solution:

$$V(2)_N = Ae^{-200t} \text{ volts}$$

This equation describes a decaying exponential voltage. The **natural response** will die out after a certain time and V(2) will closely approximate its **forced response**. The duration of the **natural response** is calculated from the **RC** product of the circuit. That product is the time constant τ of the circuit, and its value is calculated at 5 milliseconds. We note that τ is the reciprocal of the absolute value of **s**.

It remains to evaluate the constant A. The sum of the **forced** and the **natural response** of V(2), together with the initial voltage of 50 volts of the capacitor, must satisfy the differential equation for V(2) for all t > 0 seconds. From the results obtained thus far:

$$V(2) = 46.5 \sin(\omega t - 62.2°) + Ae^{-200t} \text{ volts, } t > 0 \text{ sec}$$

$$\text{At } t = 0 \text{ seconds: } V(2)_{0-} = V(2)_{0+} = 50 \text{ volts}$$

The subscripts refer to an instance before 0 seconds and an instance after t = 0 seconds. We recall that a capacitor's voltage cannot change in zero time. Using this fact we can write:

$$50 = 46.5 \sin(-62.2°) + A \text{ volts}$$

Solving this expression for A: A = 50 + 40.8 = 90.8

Thus, the **complete response** of V(**2**) for $t > 0$ seconds is equal to

$$V(2) = 46.2 \sin(\omega t - 62.2°) + 90.8 e^{-200t} \text{ volts}$$

Finally, the value of ω is equal to $2\pi f = 2(3.14)(60) = 377$ radians/second.

We compare the voltage V(2) arrived at by analysis with its trace in **PROBE**. In both cases, the capacitor voltage V(2) has an initial voltage of 50 volts. **The transient response** of that voltage decays within 25 milliseconds as shown on its trace. This time corresponds to five time constants. The calculated **transient response** is equal to five time constants. We recall that the time constant τ was equal to 5 milliseconds. Thus five time constants amount to 25 milliseconds. Thus, the two **transient responses** are of equal duration.

The **TD** of 2.88 milliseconds as measured by the **PROBE** cursors corresponds to a phase lag of V(2) of 62.2° relative to V(1). This phase shift is exactly that calculated in our analysis. It appears that both the **PSpice** program and the author agree.

Analysis of an RLC Circuit: The Overdamped Response

We shall extent our analysis to the **total response** of an **RLC** circuit. Depending upon the frequency of the applied current or voltage source, and the value of the circuit elements, the **natural, or transient, response** will either decay or break into sustained oscillations. The latter condition is possible only in circuits containing both capacitors and inductors. We shall analyze the circuit in Figure 6.02 . The resistance of **R1** set is to 300 Ω. **C1** has an initial voltage of 20 volts.

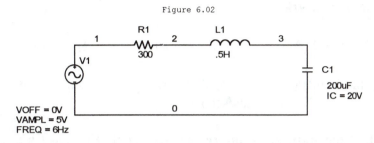

Figure 6.02

Running a **transient analysis** for 500 milliseconds produced the **PROBE** plot shown.

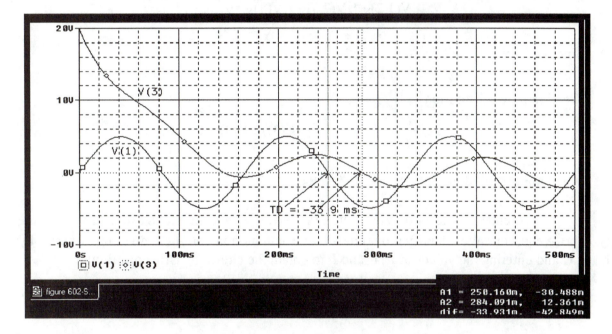

The **PROBE** plot shows the traces of V(1), essentially the source voltage, and that of V(3), the trace of the capacitor voltage. The latter decays to its steady-state value by about 250 milliseconds while it continues to lag V(1) by a **TD** of 33.9 milliseconds, which computes to a 69.5° phase lag. Let us analyze the response of this circuit and compare the results with their **PROBE** traces

We start with determining the **forced response**. Finding the phasor current $\mathbf{I_C}$, which is the unique current flowing in the circuit, we obtain:

$$\mathbf{I_C} = \frac{V(1)}{(300 + j18.8 - j133)} = \frac{3.54\angle 0°}{320\angle -20.8°} = 11\angle 20.8° \;\; \text{mA}$$

$$R = 300 \ \Omega, \ X_C = 133 \ \Omega, \ \text{and} \ X_L = 18.8 \ \Omega$$

The **forced response** of V(3) is:

$$V(3)_F = \mathbf{I}_C{}^*(jX_C) = (11\angle 20.8 \ \text{mA})(133\angle{-90°} \ \Omega) = 1.5\angle{-69.5°} \ \text{V}$$

The associated time function is $V(3)_F = 2.1\sin(\omega t - 69.5°)$ volts. In this equation, $\omega = 2\pi f = 2(3.14)(6) = 37.68$ radians/second.

To obtain the **natural response**, we start with Kirchhoff's voltage law for the source free circuit.

$$V(1,2)_N + V(2,3)_N + V(3)_N = 0 \ \text{volts}$$

Since $I(C) = C\dfrac{d}{dt}(V(3))$ amps

We obtain by substitution:

$$LC\frac{d^2}{dt^2}(V(3)) + RC\frac{d}{dt}(V(3)) + V(3) = 0 \ \text{volts}$$

We assume a solution for $V(3)_N = Ae^{st}$ volts.

We enter this form of $V(3)_N$, its first derivative and its second derivative into the above equation. We simplify it by removing the common factor Ae^{st}, dividing all terms by **LC** and entering the values of **R, C** and **L** to obtain the **characteristic equation**:

$$s^2 + \frac{R}{L}s + \frac{1}{LC} = s^2 + 600s + 10{,}000 = 0$$

We solve for **s** by using the quadratic formula. The term **1/LC** is the square of the undamped **natural frequency** of the circuit. It is that frequency at which the circuit will oscillate if all resistance is removed from it. More of this later. Solving for the two values of **s** (there are two because this is a second degree equation), we get:

$$s_1 = -\frac{R}{2L} + \left[\left(\frac{R}{2L}\right)^2 - \frac{1}{LC}\right]^{\frac{1}{2}}$$

$$s_2 = -\frac{R}{2L} - \left[\left(\frac{R}{2L}\right)^2 - \frac{1}{LC}\right]^{\frac{1}{2}}$$

For both values of **s**, the quantities within the brackets are termed the **discriminant**. Its value determines one of three possible responses of the circuit. Setting the resistance of **R1** to 300 Ω makes the **discriminant** positive. In that case, s_1 and s_2 are both negative, real and unequal. The **natural response** will be non-oscillatory, or overdamped. Solving for **s** yields

$$s_1 = -17.2 \quad \text{and} \quad s_2 = -583$$

Hence our solution for $V(3)_N$ assumes this form:

$$V(3)_N = A1e^{-17.2t} + A2e^{-583t} > 0 \text{ seconds volts,} \quad t > 0 \text{ seconds}$$

Since the circuit contains two independent energy-storing devices, two arbitrary constants, A1 and A2, must be evaluated.

In terms of the absolute value of the time constants, we have

$$\tau_1 = 1/s_1 = 1/17.2 = 58.2 \text{ milliseconds}$$

$$\tau_2 = 1/s_2 = 1/583 = 1.72 \text{ milliseconds}$$

The time constant τ_2 causes a rapid decay of the second term involving the constant A2 compared to the first time containing the constant A1. We can neglect it. Thus, our total response for $V(3)$ can be written:

$$V(3) = V(3)_F + V(3)_N = 2.1 \sin(\omega t - 69.5°) + A1e^{-17.2t} \text{ volts}$$

From our condition of continuity of $V(3)$, we have

$$V(3)_{0-} = V(3)_{0+} = 20 \text{ volts}$$

Therefore, at $t = 0_+$ seconds,

$$V(3)_{0+} = 20 = 2.1\sin(0-69.5°) + A1 \text{ volts;} \quad A = 20 - 2.1(-.937) = 22$$

The complete response of $V(3)$ is

$$V(3) = 2.1 \sin(\omega t - 69.5°) + 22 e^{-17.2t} \text{ volts,} \quad t > 0 \text{ seconds}$$

Let us compare this result with the **PROBE** plot of V(3). The trace of V(3) starts at an initial value of 20 volts. That is also the value for V(3) if t = zero seconds is inserted into the above equation. Our time constant calculated was 58.2 milliseconds. Thus, we would expect that V(3) settles into its **steady state** by about five time constants equal to 291 milliseconds. The trace of V(3) reaches **steady-state** at about 300 milliseconds. The indicated **TD** of –33.9 milliseconds, which calculated into a phase lag of 69.5°, is identical to that obtained by our analysis. It appears that all is working out just fine!

Analysis of an RLC Circuit: The Critically Damped Response

We modify our circuit in Figure 6.02 by changing R(1) to 100 Ω. The initial voltage on **C1** is still 20 volts. The modified circuit is shown as Figure 6.03.

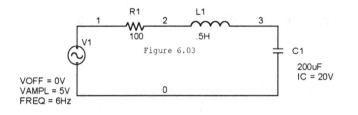

A 500 millisecond **transient analysis** obtained the traces of V(1) and V(3) on the **PROBE** plot.

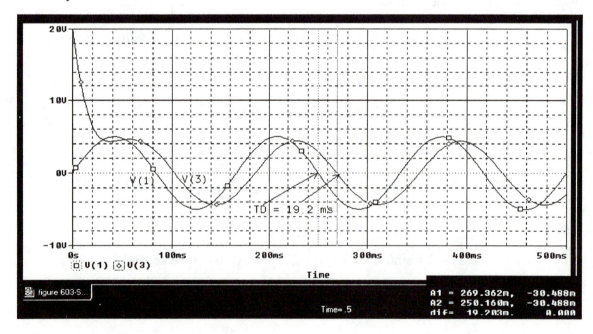

The **PROBE** plot shows the traces of V(1) and of V(3). The latter decays quickly from its initial voltage of 20 volts by contrast with the previous case. Thereafter, V(3)

closely resembles the source voltage V(1). The indicated **TD** of 19.2 milliseconds corresponds to a 41.6° phase lag.

Calculating the **forced response** with R1 set to 100 Ω, using phasor analysis:

$$\mathbf{I_C} = \frac{V(1)}{(100 + j18.8 - j133)} = \frac{3.54\angle 0°}{151\angle -48.7°} = 30.8\angle 48.7 \text{ mA}$$

$$V(3)_F = \mathbf{I_C}*(jX_C) = (30.8\angle 48.7° \text{ mA})(133\angle -90°) = 4.1\angle -41.4° \text{ V}$$

The associated time function is: $V(3)_F = 5.8 \sin(\omega t - 41.4°)$ volts, where $\omega = 37.68$ radians/second/second.

The initial steps in obtaining the **natural response** for the condition that R = 100 Ω is identical to the **overdamped** case. What changes now is the characteristic **equation.**

It reads: $s^2 + 200s + 10{,}000 = 0$

This equation has a **discriminant** of zero. Both s_1 and s_2 are negative and equal, as shown.

$$s_1 = s_2 = -100$$

This value of **s** makes the two circuit time constants $\tau = 1/s$ equal to 10 milliseconds.

The **natural response** is on the dividing line between the **overdamped** and the **underdamped response**. The value of the resistance for this condition is termed the **critical resistance**, rather aptly named.

Since for this condition, the value of the **discriminant** is equal to 0, we have

$$\frac{R^2}{4L^2} - \frac{1}{LC} = 0$$

Solving for the value of the critical resistance, we have

$$R_{critical} = 2\sqrt{\frac{L}{C}} = 100\Omega$$

The form of the solution for the **natural response** of $V(3)_N$ will be

$$V(3)_N = A1e^{st} + A2te^{st} \text{ volts}$$

The **total response** will be:

$$V(3) = V(3)F + V(3)_N = 5.8\sin(\omega t - 41.4°) + A1e^{st} + A2te^{st} \quad \text{volts}$$

To evaluate the constants A1 and A2, we use the boundary condition t = 0 seconds.

At t = 0$_+$, $20 = 5.8\sin(0 - 41.4°) + A1 + 0$ volts

Solving for A1: $A1 = 20 - 5.8(-.66) = 23.8$

We need next to solve for A2. To obtain it, we use a derived initial condition. Since there was no initial current $I(L1)$; therefore, $I(L1)_{0-} = I(L1)_{0+} = 0$ amps. Since

$$I = C\frac{d}{dt}(V(3)), \text{ and solving this equation at } t = 0 \text{ seconds, we have:}$$

$$\frac{d}{dt}(V(3)) = \frac{I}{C} = 0 \text{ volts/second}$$

We differentiate our equation for V(3) as stated above obtaining

$$\frac{d}{dt}(V(3)) = 5.8\omega\cos(\omega t - 41.4°) + sA1e^{st} + A2e^{st} + sA2te^{st} \quad \text{volts/second}$$

$$\omega = 2\pi f = 2(3.14)(6) = 37.7 \text{ radians/second}$$

at t= 0$_+$ $0 = 5.8*37.7*\cos(-41.4°) - 100*23.8 + A2$

Solving for A2: $A2 = 100*23.8 - 5.8*37.7*(.75) = 2216$

Thus, our complete expression for V(3) is

$$V(3) = 5.8\sin(\omega t - 41.1°) + 23.8e^{-100t} + 2216te^{-100t} \quad \text{volts}$$

Let us compare this result with the **PROBE** plot of V(3). The trace of V(3) starts at an initial value of 20 volts. Entering $t = 0$ seconds into the above equation will yield the sane value of 20 volts for V(3). The time constants calculated were 10 milliseconds. Thus, we would expect that V(3) settles into its **steady state** by about five time constants, equal to 50 milliseconds. The trace of V(3) reaches steady-state by about 50 milliseconds. We observe that by comparison with the **over damped case**, V(3) settles into its **steady-state** response much sooner. The indicated **TD** of -19.2 milliseconds, which corresponds to a phase lag of -41.6,° is reasonably close to the calculated phase lag of –41.4°

Analysis of an RLC Circuit: The Underdamped Response

To obtain this response, we again modify our circuit of Figure 6.02 by changing its resistance **R1** to 10 Ω. The initial voltage on **C1** remains at 20 volts. **V1** and its parameters remain the same. The modified circuit is shown as Figure 6.04. Our objective is again to obtain the capacitor voltage V(3) for t > 0 seconds.

Figure 6.04

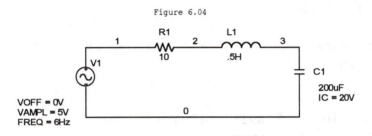

A **transient analysis** of 500 milliseconds produced the traces of V(1) and V(3). For clarity's sake, these voltages are referenced to different Y-axes. The initial oscillations of V(3) are of much greater amplitude than the 5 volt amplitude of the source voltage V(1). Because of the amplitude of V(3) and the scale of its Y-axis, the initial value of V(3) is hard to notice, although a careful look will disclose that V(3) begins above 0 volts. What is characteristic of the **underdamped response** is that it takes a long time for V(3) to settle in its steady-state mode.

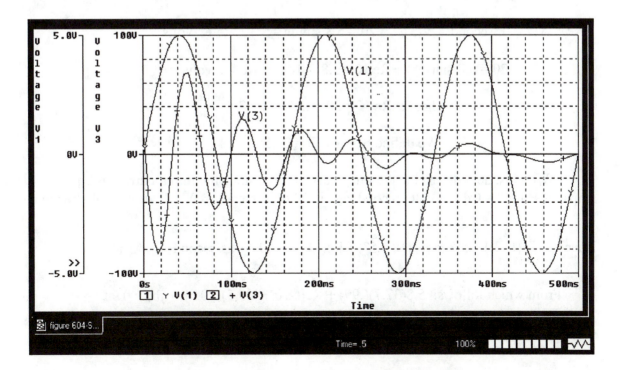

The **forced response** is obtained from phasor analysis. Using the voltage divider method, this time we get:

$$V(3)F = \frac{Xc}{(R + jXl - jXc)}V1 = \frac{(0 - j133)}{(10 + j18.8 - j133)}(3.5 + j0) = (4.1 - j.04) = 4.1\angle -5° \text{ V}$$

The corresponding time function is: $V(3)_F = 5.8 \sin(\omega t - 5°)$ volts. The radian frequency ω is still 37.68 radians/second.

To obtain the **natural response**, we enter our new value of R1 = 10 Ω into our characteristic equation and get:

$$s^2 + 20s + 10,000 = 0$$

From which: $s_1 = -10 + j99.5$ and $s_2 = -10 - j99.5$. These complex roots results in an **underdamped response** for our circuit.

The solution of the **natural response** is most conveniently put into the following form:

$$V(3)_N = \text{Re } \mathbf{K}e^{st} \text{ volts}$$

This form of the solution is preferable to the one that contains two arbitrary constants. In the present case, the arbitrary constants **A1** and **A2** must be complex conjugates because of s_1 and s_2. Since $\mathbf{A_1} + \mathbf{A_1}^* = 2\text{Re } \mathbf{A_1}$, we define $\mathbf{K} = 2\mathbf{A_1}$.

Summing up results thus far: $V(3) = 5.8 \sin(\omega t - 5°) + \text{Re } \mathbf{K}e^{st}$ volts, $t > 0$ seconds

At $t = 0_+$ seconds: $V(3)_{0+} = 20 = 5.8 \sin(\omega t - 5°) + \text{Re } \mathbf{K}$ volts

$$20 = 5.8 \sin(-5°) + \text{Re } \mathbf{K} = 5.8 (-.10) + \text{Re } \mathbf{K} \text{ volts}$$

hence: $\text{Re } \mathbf{K} = 20 + .58 = 20.58$

We define $\mathbf{K} = a + jb = 20.58 + jb$

We must next determine the imaginary part b. Since the initial current is equal to 0 amps, $I(0+) = 0 = Cdv/dt$ volts/second, and it follows that at $t = 0_+$, $dv/dt = 0$ volts/second.

Hence: $\frac{d}{dt}(V(3)) = 5.8 \omega \cos(\omega t - 5°) + \text{Re}(sKe^{st}) = 0$ volts/second   at $t = 0_+$

From which follows: $5.8(37.7)(.99) + \text{Re}[(-10 + j99.5)(20.58 + jb)] = 0$ at $t = 0_+$

We need concern ourselves only with the real part of **K** in the expression for V(3).

Hence, $216.5 + \text{Re}[(-10 + j99.5)(20.58 + jb)] = 216.5 + (-205.8 - 99.5b) = 0$

Solving for b: $b = \dfrac{(-216.5 + 205.5)}{99.5} = .11$

From which **K** = $20.58 - j.11 = 20.58\angle.3°$

We enter this value of **K**, a quantity very close to a real number, into our expression for V(3). Thus,

$$V(3) = 5.8\sin(\omega t - 5°) + 20.58e^{-10t}\cos(99.5t - .3°) \text{ volts}$$

V(3) consists of two terms. The first is a sinusoid with an amplitude of 5.8 volts. It is the **steady-state or forced, response** and will persists as long as **V1** is applied. The second term is the **free,** or **natural, response** of V(3). It consists of a decaying exponential factor and a cosine function. The exponential function provides the envelope within which the cosine function can oscillate. As that exponential factor decays, so will the oscillation of the cosine function. This is clearly indicated by the **PROBE plot** of V(3).

RESONANCE IN ELECTRICAL CIRCUITS

The **total response** of V(3) has two frequencies associated with it. One was the frequency of **V1** of 6 Hz, or 37.7 radians/second. It determines the frequency V(3) after the **transient response** is over. The second frequency, 99.5 radians/second, or 15.84 Hz, was the frequency of the **natural response** of the circuit. It has the name of **damped natural frequency**. This latter frequency is a property of the circuit independent of the frequency of any applied source. The question arises: What if the frequency of the source, such as **V1**, is equal to that **damped natural frequency?**

To answer that question, let us modify the circuit in Figure 6.04 by changing the frequency of **V1** to 15.84 Hz. This modified circuit is shown as Figure 6.05.

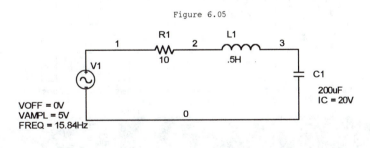

Figure 6.05

We obtain a **PROBE** plot of V(3) versus the source voltage **V1** or its nodal equivalent of V(1) by performing a **transient analysis** for 300 milliseconds.

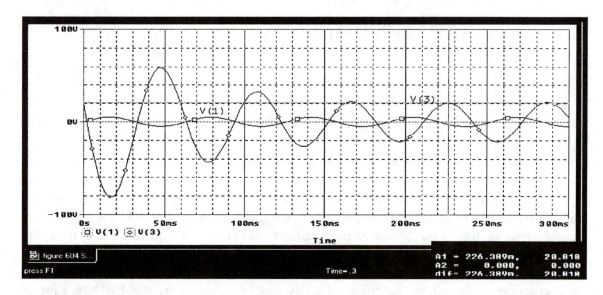

The trace of V(3) reveals that after a **transient response** lasting about 170 milliseconds, V(3) maintains an oscillatory response with a peak value of 20.8 volts. This is about four times the amplitude of the source voltage **V1**. There is no other frequency at which V(3) would obtain this amplitude. We next obtain a **PROBE** plot of the traces of the inductor voltage V(2,3), V(3), V(1) and V(1,2).

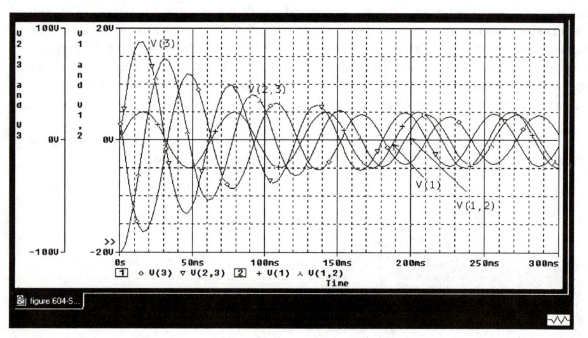

The inductor voltage V(2,3) and the capacitor voltage V(3) are equal in magnitude and have a 180° relative phase shift. If a voltmeter were placed across both elements, it would read zero volts since their relative phase shift cancels out the sum of their voltages. This in turn forces the resistor voltage V(1,2) to be equal to the source voltage **V1**. The preceding **PROBE** plot shows this to be is the case.

The equal magnitude of the reactive voltages V(1,2) and V(3) and their relative phase shift of 180° reduce the input impedance of the circuit to its minimum value equal to the 10 Ω of the resistor. At no other applied frequency will it be any smaller. The effective circuit at resonance, from the perspective of **V1**, appears as in Figure 6.06.

Figure 6.06

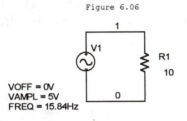

The **PROBE** plot shows the **steady-state resonance condition** for our circuit. The amplitude of the current I(R1) is solely determined by the ratio of the amplitude of the source voltage **V1**, or its nodal equivalent V(1) divided by the 10 Ω resistance of R1. Its amplitude of .5 amps is largest at the resonance frequency of 15.84 Hz.

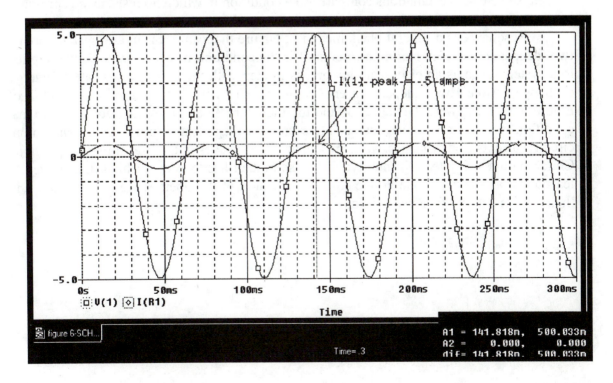

To more fully understand the performance of this circuit, we recall the **characteristic equation:**

$$s^2 + \frac{R}{L}s + \frac{1}{LC} = 0$$

The last term in this equation was previously defined as the square of the **undamped natural frequency** of the circuit. It is that frequency at which the circuit would resonate if no resistance was present in the circuit. Putting in our values for **C** and **L**, that term computes to 99.5 radians/second (Does this number look familiar?) or 15.92 Hz. However, we do have an **R1** with a resistance of 10 Ω in our circuit. To obtain the **damped natural frequency,** that is the frequency at which the circuit resonates in the presence of resistance, we solve the following equation:

$$\omega(damped) = \frac{1}{\sqrt{LC}}(1 - \frac{R^2}{4} * \frac{C}{L})$$

Entering our element values and performing the indicated operations give a calculated value of 99.48 radians/second or 15.84 Hz for the **damped natural frequency.** This shows the general result that the effect of resistance in an **underdamped RLC** circuit is always to slow its oscillations compared to a condition in which no resistance is present.

Resonance is a general phenomenon of Nature that can occur in any system having two independent energy-storing elements. It is the condition in which the energy supplied to a system, such as an electrical network, is in step with the spontaneous rate of energy exchange between two independent energy-storing elements. In the case of our circuit of Figure 6.05, these were the capacitor and the inductor. In an electrical network, it is that condition in which a stimulus such as a current or voltage source produces the maximum circuit currents and voltages. It is resonance that makes radio transmission possible. Children on a swing learn it intuitively, or else mother has to keep pushing that swing. At least as far as we know, one bridge collapsed because of it. Opera singers who attain it in their throats became famous, and rich!

PROBLEMS

6.1   For this circuit:
       (a). Find a **PROBE** trace of V(3) for 660 milliseconds starting at t = 0 seconds.
       (b). From it, determine its amplitude and phase shift relative to V(1) at **steady state**.
       (c). Estimate the length of the **Transient response** of V(3).
       (d). What is the damping of this circuit?
       (e). Determine the **resonance frequency** of this circuit.

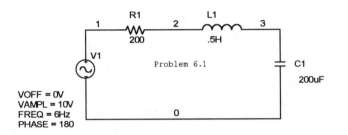

6.2   For this circuit:
       (a). Find a **PROBE** trace of V(2) for 800 milliseconds starting at $t = 0$ seconds.
       (b). From it, determine its amplitude and phase shift relative to V(1) at **steady state.**
       (c). Estimate the length of the **natural response** of V(2).
       (d). What is the **damping mode** of this circuit?
       (e). Calculate the time constant of this circuit and compare it to the length of your measured **natural response** from the **PROBE** plot.

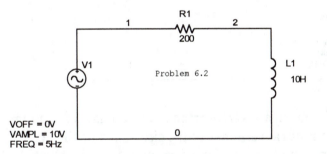

6.3   For this circuit:
       (a). From a **PROBE** trace of I(L1), determine the duration of its **natural response**.
       (b). What is its **time constant**?
       (c). What is its **AVG** value? Use the **AVG** function to determine it.
       (d). What is the peak value of I(L1)?
       (e). What is the **RMS** value of I(L1)?
       (f). Determine the **forced response** for I(L1) using phasor analysis.
       (g). From it, what is the peak-peak value of I(L1)?

(h). Compare it with that obtained from its **PROBE** trace.

(i). Determine the **free response** of I(L1).

(j). How does the calculated time constant compare with the one determined from the **PROBE** plot?

(k). Make a trace of I(L1) and I(**V1**) on the same **PROBE** plot.

(l). How do they compare?

(m). What is the phase shift of the **forced response** of I(L1) versus V(2)?

(n). What is the phase shift of the **forced response** of I(R1) versus V(1,2)?

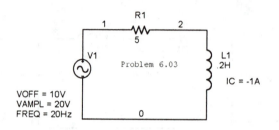

6.4 For this circuit, find the amplitudes of the currents I(L1) and I(L2). What is the relative phase shift between these two currents? Determine the duration of the **natural response** and from it determine the time constant of the circuit. From that time constant, determine the effective inductance in this circuit. Why was the resistor **Rloop** inserted into the circuit?

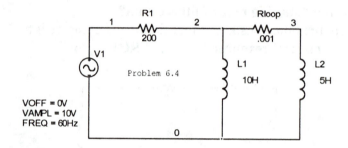

6.5 For this circuit, obtain a **PROBE** trace of the voltage V(2). From it determine the duration of the **natural response** of the circuit and its time constant. Plot a phasor diagram and from it obtain the **forced response** of the voltage V(2). Compare the calculated response of that voltage with its trace on the **PROBE** plot.

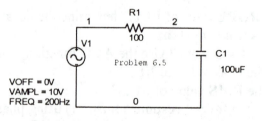

6.6   The circuit in Problem 6.5 has a capacitor C2 added to it, as shown in Problem 6.6.
        For this circuit, compare the voltage V(2) in the present and the previous problem.
        Answer all questions as in the previous problem and compare the results. Also,
        what is the effective capacitance in Problem 6.6? From it design a rule to
        determine the effective capacitance of two capacitors in parallel.

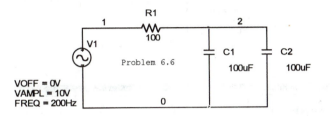

6.7   The circuit in this problem has no initial energy storage.
        (a). Obtain a trace of the capacitor voltage V(3).
        (b). Determine the length of the **natural response.**
        (c). What is the time constant of the circuit?
        (d). Obtain the **natural response** of the circuit. Hint: Set voltage source **V1** to zero
              volts and put an initial voltage of 20 volts across the capacitor, run an analysis
              with this condition and obtain the traces of the circuit voltages V(1,2), V(2)
              and V(3).
        (e). From your data just obtained, what is the form of the **natural response** of this
              circuit?

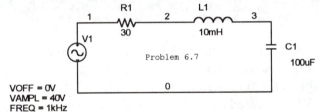

6.8   The circuit in this problem has an initial current of .5 amps flowing. Perform a
        PSpice analysis.
        (a). Obtain a **PROBE** trace of the current I(L1).
        (b). From it, determine the length of the **natural response** and the time constant of
              the circuit.
        (c). What is the form of the **natural response**?
        (d). Compare its value with the ratio of L(1)/R(1). Are they the same?
        (e). Calculate the **forced response** of this circuit.
        (f). From it, what is the peak value of the current I(L1)?
        (g). How does the calculated value of that current compare with its trace?

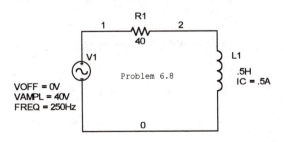

6.9   The current I(L1) has the **natural response** shown on the **PROBE** plot.

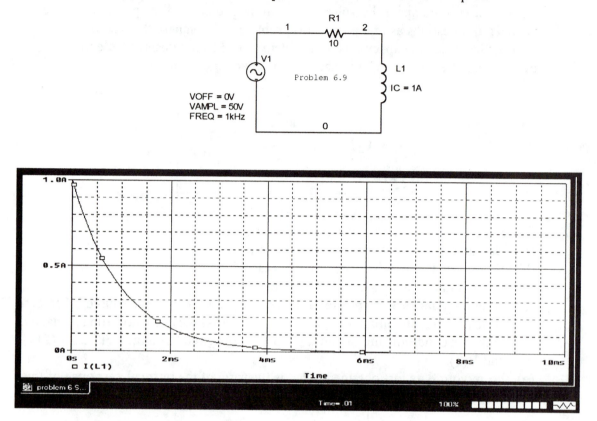

From that data, find the inductance of L1. Verify the correctness of your findings
by running a **PSpice** analysis that will give the same **natural response.** Calculate
the **forced** and the **natural responses** for the variable I(L1). Run a **PSpice** analysis
with the source **V1** connected to the circuit and verify your calculations.

6.10   For this circuit, start by obtaining the **natural response** of the circuit. This is easily
obtained by setting the voltage of **V1** equal to zero volts. Run a **transient analysis**
of at least 5 milliseconds. Plot voltage V(2) and the current I(L1) on two different
Y-axes. From their traces, determine the length of the **natural response** and the
time constant of the circuit. Do both of these variables have the same exponential
rate of change? What is the nature of the damping in this circuit? Next, change the
amplitude of **V1** to 10 volts. Run the analysis and obtain a trace of V(2). How long
does it take for that voltage to settle in its steady state? How does that time
compare with the length of the **natural response** obtained previously? Compare
V1, or its nodal equivalent V(1), with V(2). Do they approximate each other and at
what time?

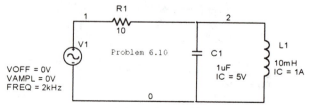

6.11   For this circuit with its initial condition, find the **natural response** of this circuit.

Problem 6.11

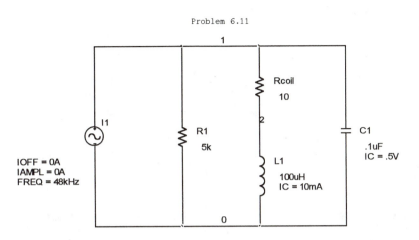

On a **PROBE** plot, obtain the traces of I(C1) and I(L1). To obtain them, run a **transient analysis** of 100 microseconds duration.   What is their relative phase shift? What kind of damping do the traces show? What is the length of the **natural response** and what is the value of their time constant? Change the amplitude of the current source I1 to 10 milliamps and run the **transient analysis** again. What are the amplitudes and the relative phase shifts of the forced components of I(C1) and I(L1)? How long does it take for them to reach **steady state**? Compare that time with the length of the **natural response** obtained above. At resonance, the two reactances of C1 and L1 are equal, thus their currents have the same magnitudes and their relative phase is 180 °. As you compare the amplitudes of the two currents I(C1) and I(L1), can you tell if the source frequency of 48 kHz is above or below the resonance frequency of this circuit.

6.12   For this circuit, the current source I1 is set to zero amps to allow for the determination of the **natural response** of this circuit.   Run a **transient analysis** of 1 microsecond duration. Plot the trace of V(2) on one Y-axis and the traces of I(C1) and I(L1) on a second Y-axis. From the data obtained, what are the extreme values attained by the three variables?   What is their approximate value after 1 microsecond?   How would you classify these waveforms? What is the relative phase shift between I(C1) and I(L1)? What is the relative phase shift between V(2) and the currents I(C1) and I(L1)? If the resistance of **Rcoil** was equal to zero ohms, what would the currents I(C1) and I(L1) look like?   To answer the last question, run the **PSpice** analysis again with **Rcoil** set to zero ohms.

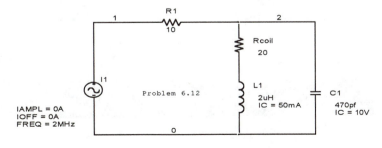

6.13   A **PSpice** analysis of the circuit in Problem 6.13 produced the two **PROBE** plots.
       The first shows the **natural response** of the inductor current I(L1). The second plot
       shows the steady-state reactance of the inductor.

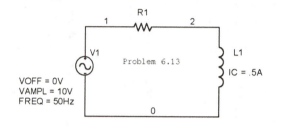

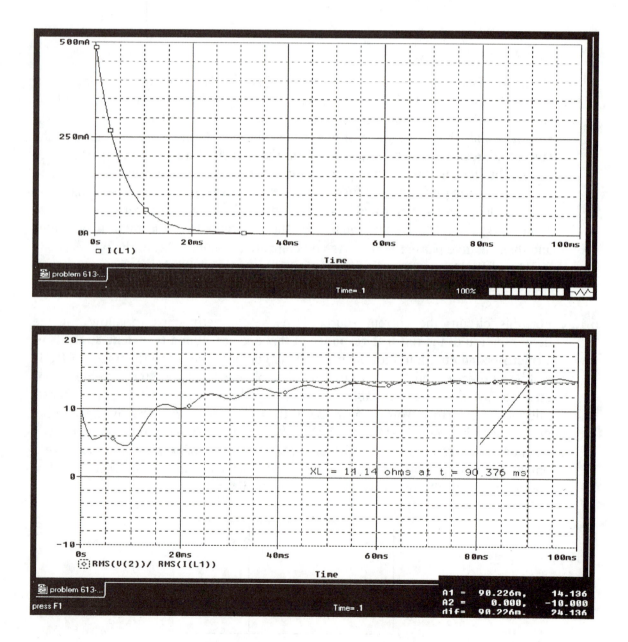

From the given information, find the value of the resistor **R1** and the inductor **L1**.
Run a **PSpice** analysis and verify the correctness of your analysis.

6.14   The **natural response** of the voltage V(3) for this circuit is shown below.

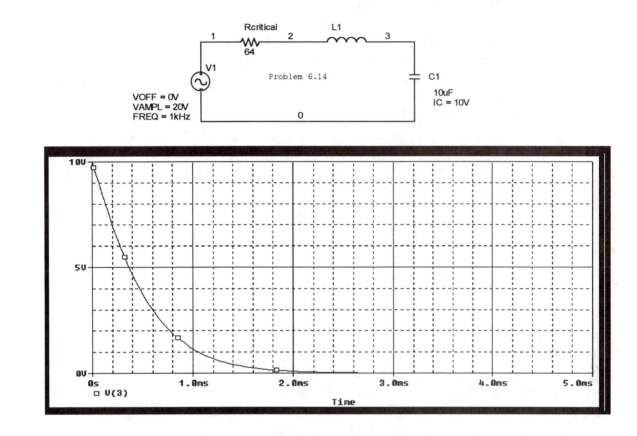

From the condition of critical damping, calculate the value of the inductance of L1. Run a
transient analysis of 5 milliseconds to determine the correctness of your calculation.
Change the amplitude of **V1** to 20 volts and repeat the **Transient analysis**. Obtain a trace
of V(3) and determine the length of its **natural response**. Is it the same as was obtained
from the above **PROBE** plot? What is the steady-state amplitude of V(3) and its phase
shift relative to V(1)? On the same **PROBE** plot, obtain a trace of the inductor voltage
V(2,3) and the capacitor voltage V(3). Which of the two voltages is larger? What is their
relative phase shift?

6.15   For this circuit, find the following:
        (a). Obtain the **natural response** by setting the voltage of **V1** equal to zero volts.
        (b). Get the trace of V(3).
        (c). What is the time length of the **natural response**?
        (d). Set the amplitude of **V1** to 10 volts and obtain the **forced response** of voltage
              V(3).
        (e). Plot the traces of V(1) and V(3).
        (f). What is the operating mode of this circuit?
        (g). How long does is take V(3) to reach **steady state**?
        (h). Compare your finding with that of the **natural response** obtained above.
        (i). Perform a phasor analysis to confirm your findings.

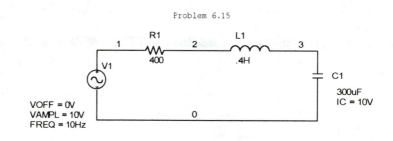

Problem 6.15

7

ALTERNATING CURRENT
NETWORK THEOREMS

INTRODUCTION

In this chapter we shall expand our applications of the various circuit theorems covered in Chapter 2 to circuits that are excited by sinusoidal currents and voltages. The circuits that will be analyzed contain resistors, capacitors and inductors. The application of Thevenin's and Norton's theorems will result in finding a particular current or voltage source together with an impedance that represents the circuit exclusive of the load. The application of these theorems results in determining the amplitude of load current and voltage and the determination of a phase shift between source and load voltage. It will be shown that the maximum power transfer between a source and a load occurs when the load impedance is the complex conjugate of the **Thevenin impedance**.

THE SUPERPOSITION THEOREM APPLIED

Conventional Analysis of Figure 7.01

The circuit in Figure 7.01 contains two independent voltage sources, **V1** and **V2**. Our objective is to find the current through and the voltage across **R3** by a conventional analysis with the two sources connected in the circuit. Next, we shall apply the superposition theorem by obtaining the load current I(R3) and voltage V(3) across **R3** due to either of the voltage sources. Their contributions will be summed to get the total current I(3) and the total voltage V(3). Finally, the results of the two different analyses will be compared to see if they yielded identical results. We shall perform a 35 millisecond **transient analysis.**

Figure 7.01

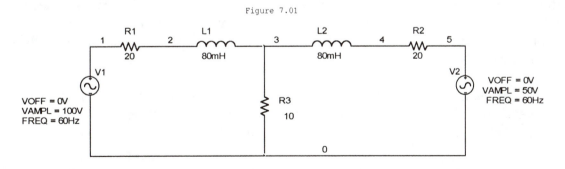

The **PROBE** plot shows the traces of the two sources voltages, **V1** and **V2** and the trace of the nodal voltage V(3). The traces of the nodal equivalents of **V1** and **V2** are shown as V(1) and V(5), respectively. Their amplitudes are 100 volts and 50 volts, with

203

no relative phase. The amplitude of V(3) is 29.9 volts and it has a **TD (time delay)** of 1.72 milliseconds relative to **V1** or its nodal equivalent V(1). That time delay corresponds to a 37.1° phase lag. Its expression is:

$$V(3) = 29.9 \sin(\omega t - 37.1°) \text{ volts}, \quad \omega = 2(3.14)(60) = 377 \text{ radians/second}$$

A reminder: to get the **TD**, activate the cursors. A small rectangle will enclose the symbol for V(1). It means that cursor **A1** will move along its trace. Move cursor A1 to indicated spot on trace V(1). Right click on V(3). A small rectangle will now enclose its symbol. Depress the Shift key together with the **Right Arrow→** key. Move cursor **A2** to indicated spot on trace V(3). In the coordinate box, **dif** = -1.72ms is the **TD.**

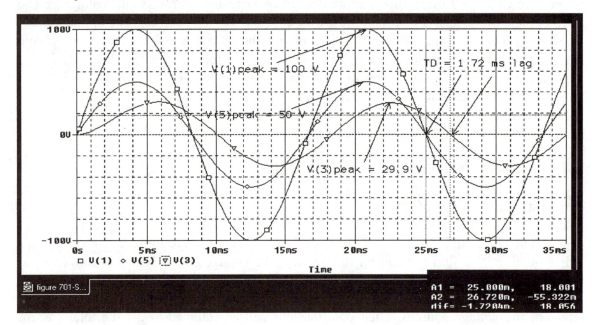

The Superposition Theorem Applied

The superposition theorem is applicable to circuits containing resistors, capacitors and inductors. Their presence does not violate the principle of linearity, a necessary condition for the theorem. We do recall however that power is nonlinear. The theorem can be applied to circuits containing both independent and dependent current or voltage sources or a combination of them.

To apply the theorem, we remove voltage source **V2** by shortening it. This is most conveniently done by setting its **VAMP** = 0 V. The resultant circuit, labeled Figure 7.02, is shown next.

Figure 7.02

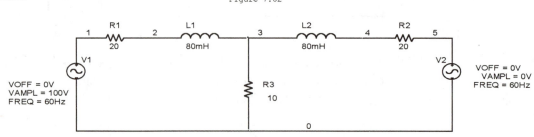

The **PROBE** plot shows the result of a 35 millisecond **transient analysis**.

The trace of V(3) has an amplitude of 19.937 volts and it lags the voltage source **V1** or its nodal equivalent V(1) by 1.7742 milliseconds. That delay computes into a 38.2° phase lag. Its mathematical expression is:

$$V(3)_{\text{due to V1}} = 20 \sin(\omega t - 38.2°) \text{ volts}, \quad \omega = 377 \text{ radians/second}$$

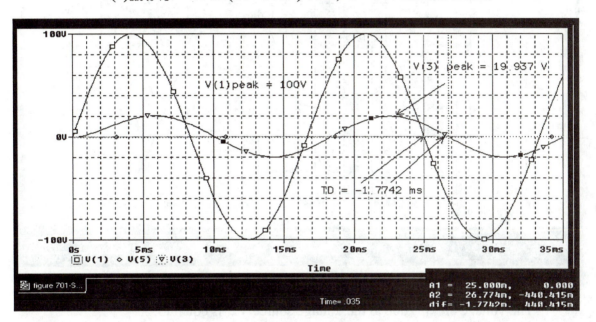

We restore the voltage of **V2** and shorten the voltage source **V1**. The modified circuit is shown as Figure 7.03.

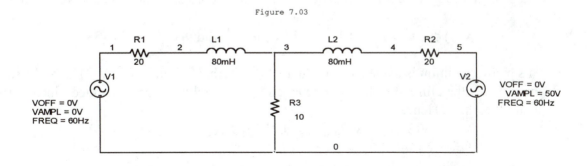

Figure 7.03

Running a **transient analysis** of 35 milliseconds of this circuit produced the **PROBE** plot shown next.

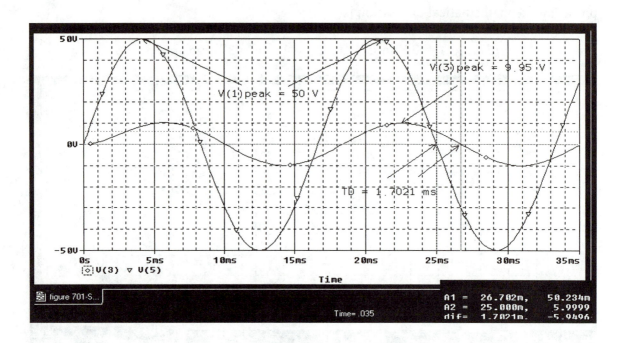

The voltage of V(3) due to the voltage source **V2** is equal to:

$$V(3)_{\text{due to V2}} = 10\sin(\omega t - 38.2°) \text{ volts}, \quad \omega = 377 \text{ radians/second}$$

If all goes well, the sum of the two components of V(3) is equal to the voltage V(3) obtained by conventional analysis. To check this out, we shall use phasor analysis.

Converting the components of V(3) into their corresponding phasor we obtain:

$$V(3) \text{ due to } \textbf{V1} = 14.1\angle\text{-38.2 °V} \text{ and } V(3) \text{ due to } \textbf{V2} = 7\angle\text{-38.2° V}$$

This phasor addition is made easy by the fact that both components of V(3) have the same phase delay; therefore, the phasors line up and their amplitudes can be added linearly to obtain their sum. Hence

$$V(3)_{\text{total}} = V(3)_{\text{due to V1}} + V(3)_{\text{due to V2}} = 21.1\angle\text{-38.2° V}$$

The corresponding time function is $V(3) = 29.8 \sin(\omega t - 38.2°)$ volts.

This voltage of V(3) compares well with that obtained by conventional analysis. The reason that the two **TDs** are equal is because both **V1** and **V2** are connected to the same combination of resistive and reactive elements. The slight difference in the phase shift is a result of the slight error introduced when the cursors are aligned to measure the **TD.**

THE THEVENIN THEOREM APPLIED

Conventional Analysis of Figure 7.04

In this section, we expand the application of **Thevenin's theorem** to include circuits that have sinusoidal sources and a combination of resistors, capacitors and inductors. The **Thevenin voltage source** will be a sine wave if the original source voltage, or sources, is a sine wave. In general, it will differ in magnitude and phase from the original source(s).

The **Thevenin impedance**, in general, will be complex. It has both a magnitude and a phase angle. Thus, it consists of a real part, a resistor and a reactive part. The latter can be a capacitor or inductor. This in turn makes the Thevenin equivalent circuit dependent upon the operating frequency of the voltage source.

Let us find the Thevenin equivalent circuit for the network of Figure 7.04 by starting with the circuit "as is." We begin by obtaining the nodal voltage V(3) across **Rload**. The frequency of the voltage source **V1** is 1kHz.

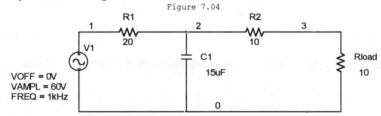

Figure 7.04

The resultant **PROBE** plot of a 2 millisecond **transient analysis** is shown next.

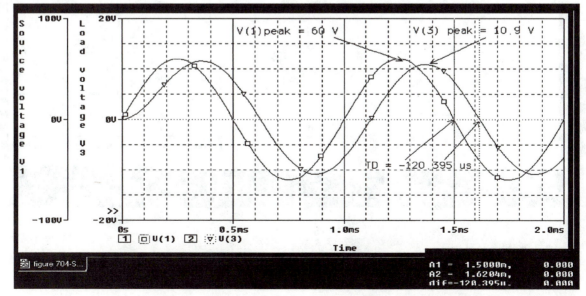

Its plot shows the trace of the load voltage V(3) has an amplitude of 10.9 volts and a **TD** of −120.395 microseconds relative to the source voltage **V1**, or its nodal equivalent V(1). This computes into a phase lag of 43.2°. The load voltage V(3) was referenced to a second Y-axis for reasons of scale. This explains the apparent similarity of the amplitudes of V(1) and V(3). Expressing the load voltage V(3) as a function of time, we obtain:

$$V(3) = 10.9 \sin(\omega t - 43.2°) \text{ volts, where } \omega = 2(3.14)(1000) = 6280 \text{ radians/second}$$

The Thevenin Theorem Applied to Figure 7.04

Determining the Thevenin Voltage

In Figure 7.04, to obtain the open-circuit **Thevenin voltage**, we remove **RLoad** by replacing it with a very large resistor. A resistance of 10 MΩ will do just fine. The reason that we simply cannot remove **RLoad** is that the syntax of the **PSpice** program does not allow for floating nodes. In our case, node 3 would be floating. We run a **PSpice** analysis with the modified circuit of Figure 7.04 shown as Figure 7.05.

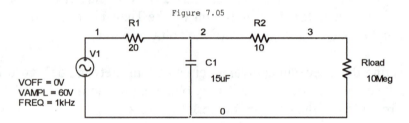

Figure 7.05

Running a **transient analysis** of this circuit produced the **PROBE** plot shown.

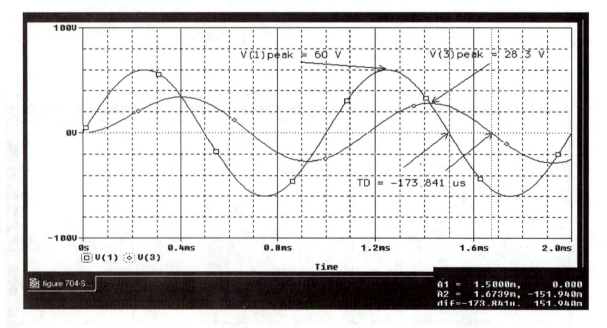

The trace of the open circuit voltage V(3) has an amplitude of 28.3 volts and a **TD** of -173.841 microseconds. This **TD** corresponds to a 62.3° phase lag relative to the source voltage **V1** or its nodal equivalent voltage V(1). Expressing the Thevenin voltage as a function of time:

$$V_{Thevenin} = 28.3 \sin(\omega t - 62.3°) \text{ volts} , \text{ where } \omega = 2(3.14)(1000) = 6280 \text{ radians/second}$$

It is important to realize that this is the open circuit voltage V(3). It is the voltage that would appear if no resistor was connected to node 3.

Determining the Thevenin Impedance

The method we use with the **PSpice** program is essentially the same method a laboratory technician would employ to determine that impedance. We begin by connecting a voltage source **VTEST** between node 3 and 0. Its amplitude is somewhat arbitrary, so 1 volt will do just fine. The voltage of **V1** must be set to zero volts. In a laboratory setting, such a voltage source would be removed and replaced with a conducting wire.

The amplitude of the **Thevenin impedance** is equal to the ratio AVG(RMS(V(3))/RMS(I(R2)). The **AVG** (average) of this ratio will be plotted to reduce the perturbations of that ratio until **steady state** is reached. If the current I(VTEST) had been used in the denominator of this ratio, the result would have been a negative amplitude because of the syntax of the **PSpice** program. The reader will recall that any current leaving the positive terminal of a voltage source is defined as a negative. Finally, **RLoad** stays disconnected from the circuit. Our modified circuit is shown in Figure 7.06 and the **PROBE** plot resulting from a **transient analysis** is shown next.

Figure 7.06

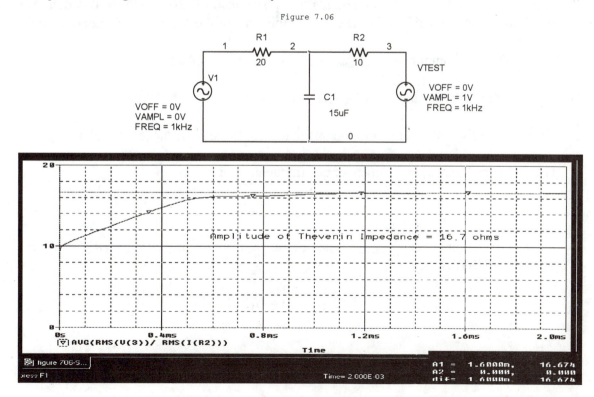

The amplitude of the **Thevenin impedance** is equal to 16.7 Ω. As was indicated before, the **Thevenin impedance** in general is a complex quantity. Thus, we need to determine the real and the reactive component of this impedance. For this, we need to determine the phase shift between the voltage source **VTEST** and the current I(R2). This was done by obtaining the traces of **VTEST** and I(R2) and determines their relative **TD**. The result is shown next.

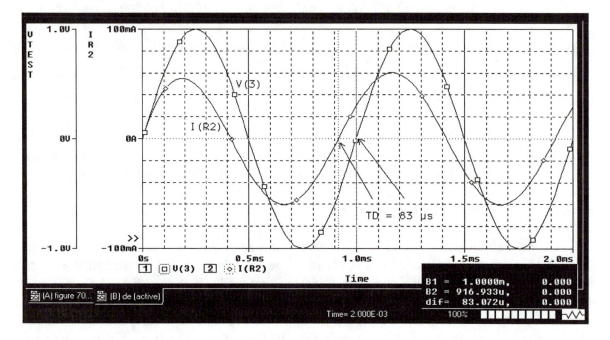

The current I(R2) has a peak value of 60 milliamps and leads the voltage **VTEST** by 83 microseconds. This corresponds to a phase lead of 30.2°. From this data, we can easily calculate the real and the reactive components of the **Thevenin's impedance**.

The real component $R_{Thevenin}$ is 16.6 cos(30°) = 14.4 Ω.

The reactive component $X_{Thevenin}$ is 16.6 sin(30°) = 8.4 Ω.

It remains to find the circuit components that yield those impedances at 1 kHz.

The real part, $R_{Thevenin}$, is independent of frequency. Therefore

$$R_{Thevenin} = 14.4 \ \Omega$$

The reactive component, $X_{Thevenin}$, must be capacitive since the current I(R2) is leading the voltage **VTEST.** Hence

$$C_{Thevenin} = 1/(2*\pi*f*X_{Thevenin}) = 1/(2*3.14*1000*8.4) = 19 \ uF$$

We can now construct the **Thevenin equivalent circuit** of Figure 7.07. Carefully note that the phase of the **Thevenin voltage source** must be entered. This is done via the Property **Editor.**

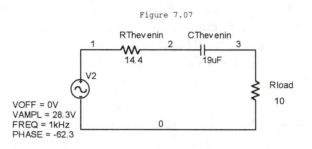

Figure 7.07

It remains to verify that the voltage V(3) across **RLoad** and the current through it are the same as for the circuit in Figure 7.04. We run a **PSpice** analysis of the circuit in Figure 7.07 with the results shown.

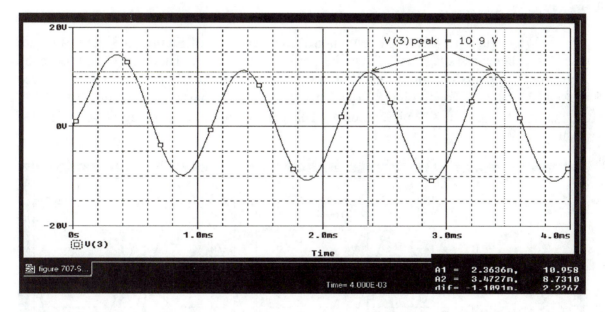

A **transient analysis** of 4 milliseconds was performed to allow V(3) to reach steady-state. The amplitude of the present load voltage V(3) of 10.9 volts is identical to that in Figure 7.04. The peaks of V(3) in Figures 7.04 and 7.07 occur at the same time. Thus, from the perspective of **RLoad**, the circuits of Figures 7.04 and 7.07 are equivalent. Should **RLoad** be changed, the circuit in Figure 7.07 would allow for a quicker determination of the new load voltage compared to Figure 7.04. This becomes especially important if such calculation were done without the benefit of the **PSpice** program.

Further Application of Thevenin's Theorem

Conventional Analysis

The circuit shown in Figure 7.08 could represent the lumped parameter model of a transmission system of many miles of length. Our objective is to find the Thevenin equivalent circuit with respect to **RLoad**. We shall also study the transmission of power from source to load and the effects of changing the applied frequency on system performance.

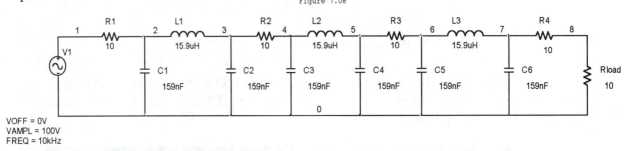

Figure 7.08

We begin our analysis by running a **PSpice** analysis of Figure 7.08. Its results are used as a benchmark against which the performance of our Thevenin circuit is to be compared. A **transient analysis** of 200 microseconds produced the traces of the source voltage **V1** and the load voltage V(8). That latter voltage has a peak voltage of 18.541 volts and a **TD** of -10.776 microseconds. This corresponds to a phase lag of 38.8° of V(3) relative to **V1**.

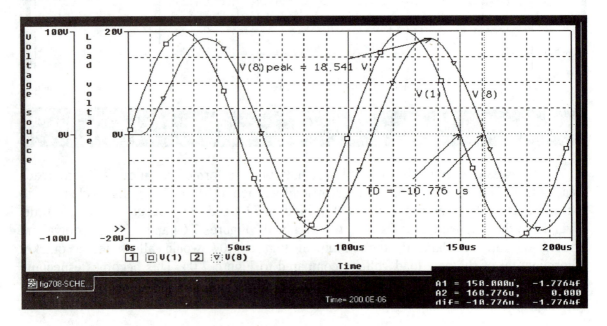

Determining the Thevenin Voltage

We replace **RLoad** with a 10 MΩ resistor and run a **PSpice** analysis to obtain the open-circuit **Thevenin voltage**. Its trace is shown next.

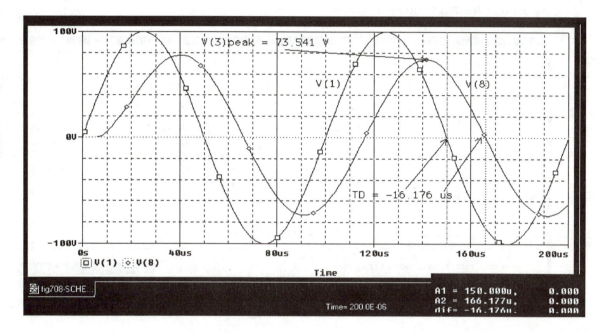

The open-circuit voltage V(3) has an amplitude of 73.541 volts and a **TD** of -16.176 microseconds. That delay translated into a 58.23° phase lag. Thus, the Thevenin voltage is:

$$V_{Thevenin} = 73.5 \sin(\omega t - 58.2°) \text{ where } \omega = 2\pi f = 2(3.14)(10 \text{ kHz}) = 62800 \text{ radians/second}$$

Determining the Thevenin Impedance

We modify the circuit in Figure 7.08 as shown in Figure 7.09.

Figure 7.09

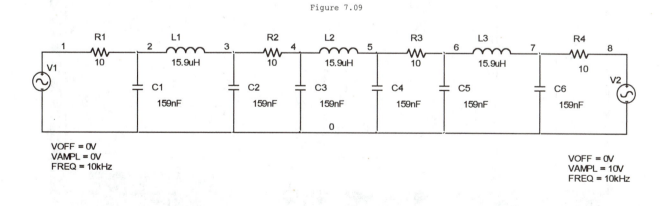

The voltage source **V1** has been shortened. A voltage **V2** has replaced **RLoad**. Its amplitude was chosen arbitrarily at 10 volts. Its frequency is the same as that of **V1**. Running a **transient analysis** of 10 milliseconds produced the traces shown. The transient time interval was chosen to insure that the **Thevenin** impedance reached its steady-state condition. Its magnitude is 30 Ω.

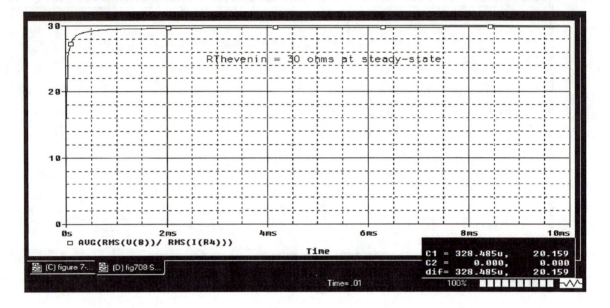

To determine the real and reactive component of this impedance, we need to obtain the phase shift between the voltage source **V2,** or its nodal equivalent V(8), and the current I(R4). The **PROBE** plot shows that I(R4) leads V(8) by 7.14 microseconds. This corresponds to a 25.7° phase lead.

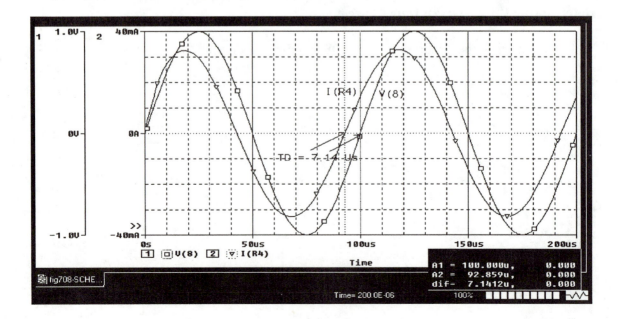

Hence

$$R_{Thevenin} = 30 \cos(25.7°) = 27 \, \Omega$$

A phase lead means that the reactive component must be capacitive.
Therefore

$$X_{Thevenin} = 30 \sin(25.7°) = 13 \, \Omega \text{ (capacitive)}$$

From which

$$C_{Thevenin} = 1.22 \text{ uF}$$

Our Thevenin equivalent circuit of Figure 7.08 is shown in Figure 7.10.

Figure 7.10

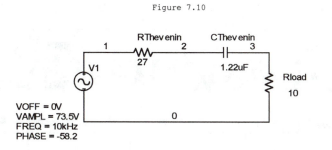

A **transient analysis** of 400 microseconds was used to allow V(3) to settle in its steady-state. Its peak voltage is 18.6 volts, which compares closely with the V(3) of 18.54 volts obtained for the circuit of Figure 7.08. It is important to notice that the peaks for both load voltages occur at the same time. By way of example: the second peak for V(3) in Figure 7.08 and in Figure 7.10 occurs at 138 microseconds. This is a must because both voltages, if all goes well, must be identical, including the times of their peaks. This is despite the fact that the original voltage source **V1** had zero phase, while the Thevenin source has a phase lag of 58.2°.

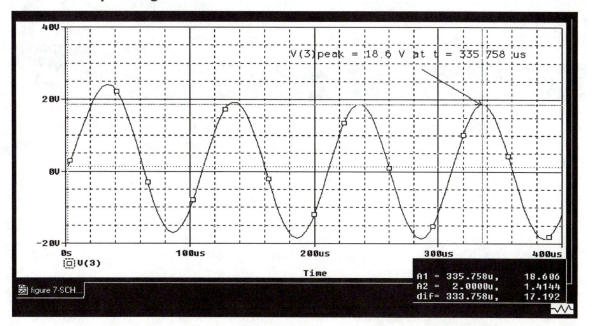

Does the equivalency between the circuits in Figures 7.08 and 7.10 hold for the power received by Rload?

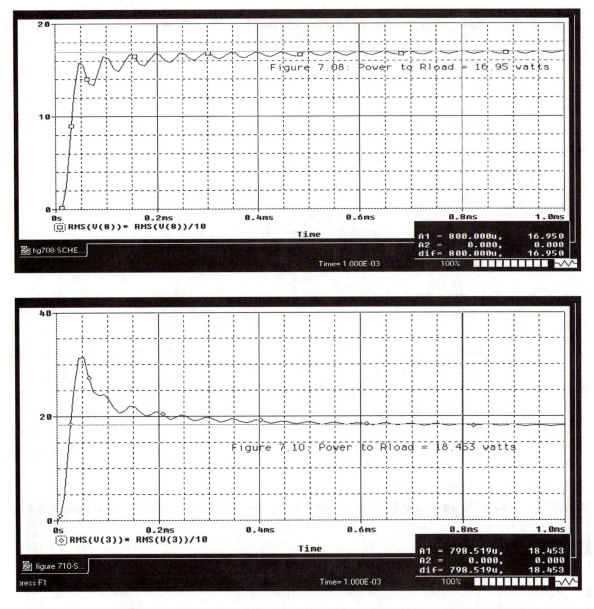

A comparison of the two traces shown above of the power delivered to **Rload** by either of the circuits in Figure 7.08 and Figure 7.10 shows them to agree within 8% of each other.

THE NORTON THEOREM APPLIED

We shall next apply Norton's theorem to the circuit in Figure 7.11. It is the same circuit as that of Figure 7.04. It will allow us to establish the equivalency between the Thevenin and the Norton theorem. It is often assumed that a Norton circuit can be found only if the circuit under analysis has only dependent and independent current sources. Such is not the case. We shall demonstrate that a Norton equivalent circuit can be found having any combination of independent and dependent current and/or voltage sources. The decision about which of the two equivalent circuits to obtain, the Thevenin or the Norton circuit, is one of convenience. It is not a theoretical necessity.

Determining the Norton Current Source

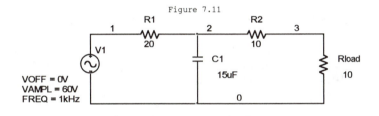

Figure 7.11

We previously determined the load voltage V(3) obtained by a conventional analysis of the circuit in Figure 7.04 was:

$$V(3) = 10.9 \sin(\omega t - 43.2°) \text{ volts}, \quad \text{where } \omega = 2(3.14)(1000) = 6280 \text{ radians/second}$$

We recall from Chapter 2 that to obtain the **Norton current source**, we need to replace **RLoad** with a short circuit. A resistance of 1 mΩ will do. Running a **transient analysis** of two milliseconds produced the two **PROBE** traces shown.

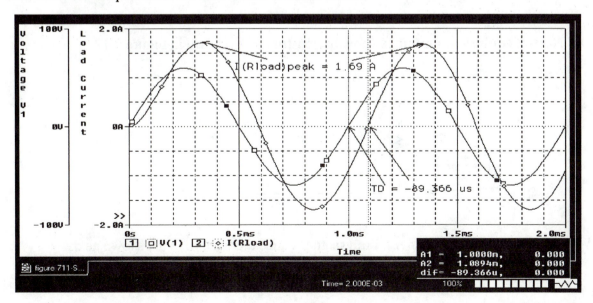

The peak amplitude of I(Rload) is equal to 1.69 amps and it lags the source voltage **V1**, or its nodal equivalent V(1) by 89.366 microseconds. This computes into a 32.2° phase lag. From this data, the **Norton current** as a function of time is:

$$I_{Norton} = 1.69 \sin(\omega t - 32.2°) \text{ amps, where } \omega = 6280 \text{ radians/second}$$

Determining the Norton Admittance

The **Norton impedance** is obtained in the same manner as the Thevenin impedance. It follows that the two impedances are the same. Yet often, the Norton source is shown as a current source with an admittance in parallel. We shall obtain that admittance. Next, we shall show the equivalency between the Thevenin and the Norton circuits.

We recall that the **Thevenin impedance** for Figure 7.04 was $14.4 - j8.8$ Ω.

To convert this into an admittance, we obtain the inverse of that impedance. Hence

$$\mathbf{Y}_{Norton} = \frac{1}{(14.4 - j8.8)} = \frac{1}{16.9\angle -31.4°} = 59.2\angle 31.4° = (50.5 + j30.8)mS$$

From this, we can obtain the needed circuit elements. There is no resistive admittance symbol. Thus, we use its inverse, a resistor of value 19.8Ω, in our **Norton equivalent circuit.**

The reactive term, defined as the susceptance, yields the needed circuit element. A positive sign indicates that the needed element is a capacitor. Its value is determined as follows:

$$C_{Norton} = \frac{30.8m}{2*3.14*1000} = 4.9uF$$

Our **Norton equivalent circuit** can now be constructed as shown in Figure 7.12.

Figure 7.12

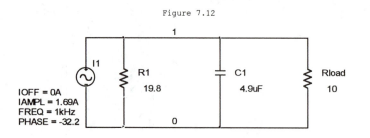
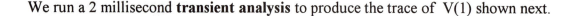

We run a 2 millisecond **transient analysis** to produce the trace of V(1) shown next.

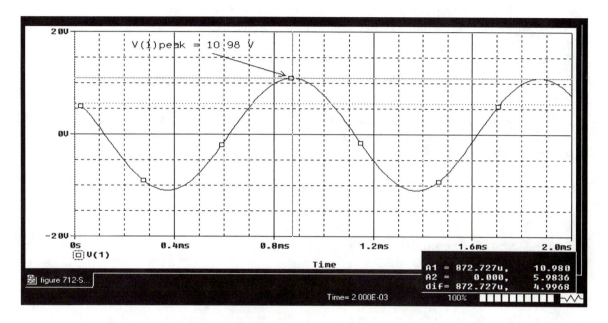

The trace of voltage V(1) across **RLoad** shown above has the identical peak value of as the voltage across the same **RLoad** in Figures 7.04 and 7.07. This proves that the **Norton equivalent circuit** of Figure 7.12 can represent both the original circuit in Figure 7.04 and its Thevenin Equivalent circuit of Figure 7.07.

The Relationship Between Thevenin and Norton Circuits

To start, we noted previously that the **Thevenin impedance** is the inverse of the **Norton admittance.** In symbols:

$$\mathbf{Z}_{\text{Thevenin}} = 1/\mathbf{Y}_{\text{Norton}} \; \Omega$$

Next, if we divide the **Thevenin phasor voltage** of Figure 7.04 by the **Thevenin impedance** of that figure, we obtain the Norton phasor current for Figure 7.12.

$$\mathbf{I}_{\text{Norton}} = \frac{V_{Thevenin}}{Z_{Thevenin}} = \frac{20\angle -62.3°}{16.67\angle -30.3°} = 1.19\angle -32° A$$

Converting the **Norton phasor current** into its corresponding time function , we get

$$I_{\text{Norton}} = (1.41)(1.19)\sin(\omega t - 32°) \text{ amps} = 1.68 \sin(\omega t - 32°) \text{ A}$$

This Norton current corresponds exactly to the previously obtained value in Figure 7.11.

Given these relationships, we can convert the **Thevenin equivalent circuit** into its corresponding **Norton equivalent circuit**. The inverse conversion also follows. The choice of either of these two circuits is often one of utility in a particular problem situation.

APPLYING THE MAXIMUM POWER TRANSFER THEOREM

We extend the application of the **maximum power transfer theorem** first applied to the resistive circuits in Chapter 2 to those circuits containing resistive and reactive elements. The statement of the theorem reads:

> *Maximum power will be delivered to a load when the load impedance*
> *is the conjugate of the Thevenin impedance of a circuit.*

Using the symbols of complex algebra to express this theorem:

$$\mathbf{Z}_{\text{load}} = \mathbf{Z^*}_{\text{Thevenin}} \ \Omega$$

The reader will recall that the * symbol denotes the conjugate of a complex number. From the perspective of the **Thevenin voltage source**, for maximum power transfer from source to load, the input impedance into the **Thevenin circuit** is:

$$\mathbf{Z}\text{in} = \mathbf{Z}_{\text{Thevenin}} + \mathbf{Z}\text{load} = \mathbf{Z}_{\text{Thevenin}} + \mathbf{Z^*}_{\text{Thevenin}} = 2 \ \text{Re} \ \mathbf{Z}_{\text{Thevenin}} \ \Omega$$

From these follows that for maximum power transfer from source to load, the input impedance is a resistor that has twice the resistance of the **Thevenin resistor**. If we compare this result with the condition of maximum power transfer of Chapter 2, we observe that in both cases the load resistance was equal to the **Thevenin resistance**. In the present case, we have one additional requirement. The impedance angle of the load must be the numerically equal to impedance angle of the source, but it must be of opposite sign. Figure 7.13 shows these conditions graphically.

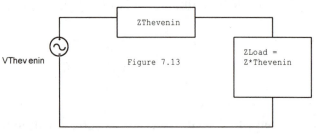

Figure 7.13

For the circuit shown in Figure 7.14, let us find the load impedance that results in maximum power transfer from source to load. We begin our analysis by obtaining the open circuit, or **Thevenin voltage**, by removing **Zload** and replacing it with a 10 MΩ resistance across the node pair 3-0.

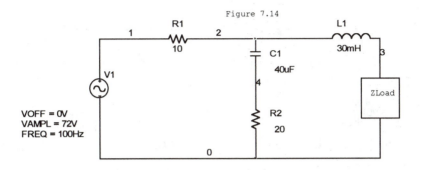

Figure 7.14

A **transient analysis** of 20 milliseconds produces the trace of the open circuit voltage V(3).

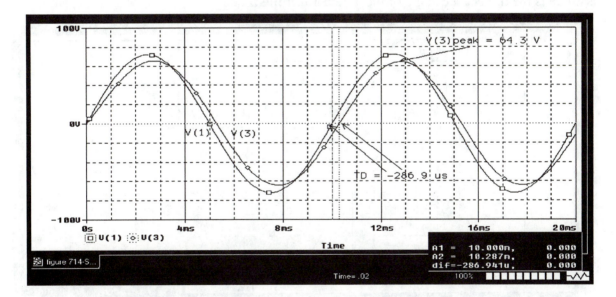

It lags the source voltage **V1,** or its nodal equivalent V(1), by 286.9 microseconds. This calculates into a 10.3° phase lag. The peak voltage of V(3) is equal to 64.3 volts. Thus, the **Thevenin voltage** is equal to:

$$V_{Thevenin} = 64.3 \sin(\omega t - 10.3°) \text{ volts, where } \omega = 2(3.14)(100) = 628 \text{ radians/second}$$

The corresponding phasor voltage is $V_{Thevenin} = 45.4\angle{-10.3°}$ V.

To obtain the **Thevenin impedance**, we modify the circuit in Figure 7.14 as shown in Figure 7.15. The voltage source **V1** has been set to zero volts. The positive terminal of **RTest** points to the positive terminal of **VTest**. This assures the correct phase shift between the voltage and the current of **VTest**.

Figure 7.15

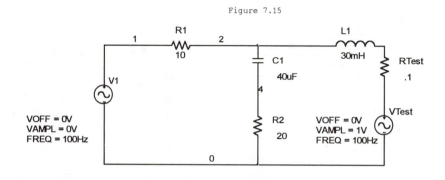

A **transient analysis** of 20 milliseconds produced the traces of **VTest** and its current. The indicated **TD** of 1.7339 milliseconds corresponds to a phase lag of 62.4° of I(RTest) relative to V(5), which is the nodal equivalent voltage of the voltage source **VTest.** This angle is the impedance angle of the **Thevenin impedance**.

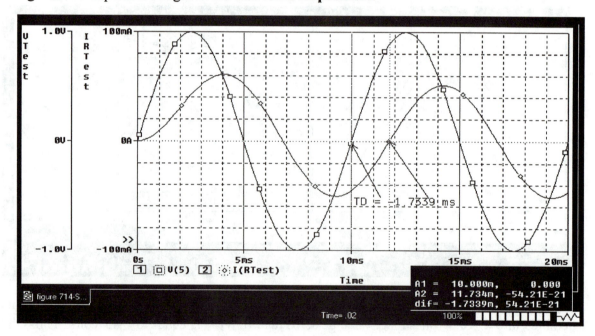

We next determine the amplitude of the **Thevenin impedance**. After steady state is reached, the amplitude of the **Thevenin impedance** is equal to 18.56 Ω, as shown on the following **PROBE** plot. This, plus the phase angle, allows us to determine the real and the reactive parts of the **Thevenin impedance**.

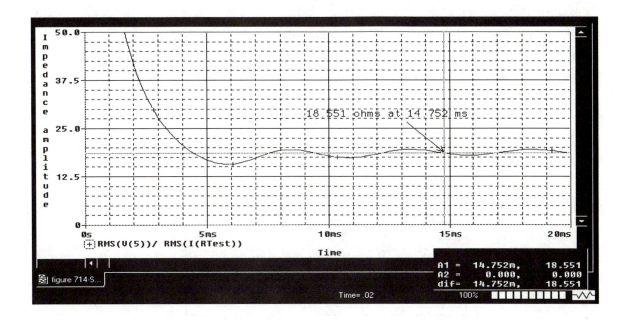

The real part is

$$R_{Thevenin} = 18.56 \cos(62.4°) = 8.6 \ \Omega$$

The reactive part must be inductive because I(RTest) lags **VTest**. Its amplitude is determined as

$$XL_{Thevenin} = 2\pi fL = 18.56 \sin(62.4°) = 16.44 \ \Omega$$

from which we determine

$$L_{Thevenin} = \ 16.44/(2*3.14*100) = 26.2 \ mH$$

We are now ready to construct the **Thevenin circuit** and obtain the maximum power to the load impedance. This is done with the circuit in Figure 7.16. The real part of our load impedance is the same as $R_{Thevenin}$. The reactive part of the load has the same reactance as $XL_{Thevenin}$, but it must be capacitive. Hence, we have to compute the capacitance of a capacitor that produces the same reactance as does $XL_{Thevenin}$.

Hence

$$XC_{Load} = \ 16.44 = 1/(2\pi fC) \ \Omega$$

From which

$$C_{Load} = 1/(2*3.14*100*16.44) = 97 \ uF$$

The completed circuit is shown in Figure 7.16, together with the traces of the load voltage V(3) and the load current I(Rload). The peak voltage of the load voltage V(3) is 69.3 volts. The load current has a peak value of 3.7 amps. The load current leads the load voltage by

1.7548 milliseconds. This corresponds to a phase lead of 63.2°. Ideally this phase lead should be numerically the same as the 62.4° phase lag of the **Thevenin impedance**. We can agree that they are reasonably close. The sum of these lead and lag angles is zero degrees: hence, from the perspective of the voltage source, the input impedance is real and is equal to

$$R_{Thevenin} + R_{Load} = 17.2 \ \Omega$$

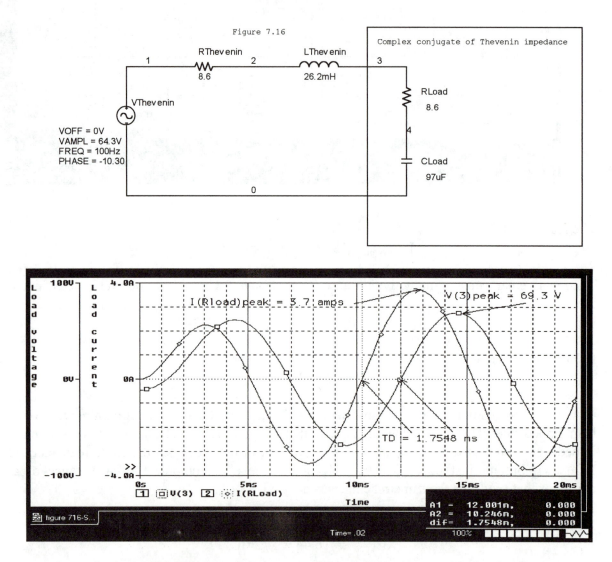

Figure 7.16

Let us obtain a **PROBE** plot of the traces of the nodal voltage V(1) that is identical to the source voltage **V1** versus I(RLoad). Since we determined that the input impedance is a pure resistor of value 17.2 Ω, we expect that V(1) and I(RLoad) are in phase and that the ratio of their peak values should be equal to 17.2 Ω. Dividing the peak voltage of V(1) by that of I(Rload) we have 64.2 volts/3.7 amps = 17.4 Ω. From the **PROBE** plot shown, our predictions prove to be true.

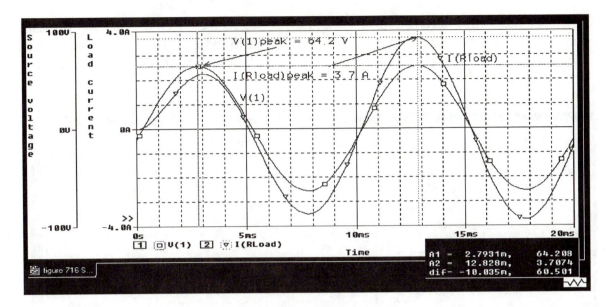

We recall from Chapter 2 that maximum power flowed from source to load when the input impedance was a pure resistor equal to twice that of the **Thevenin resistor**. This, as was just demonstrated, is also true for circuits involving complex components. Let us next compute the maximum power delivered to the load in the circuit of Figure 7.16 and verify it with a **PSpice** analysis.

For the circuit in Figure 7.16:

$$\mathbf{Z}_{in} = 2Re\,\mathbf{Z}_{Thevenin} = 17.2\Omega$$

From the **PROBE** trace of I(Rload), the phasor current **I(RLoad)** is: $2.6\angle 63.2°$ A.

Therefore
$$P_{max} = [\mathbf{I(RLoad)}]^2 * Rload = [2.6]^2 * 8.6 = 58.8 \text{ watts}$$

This is the maximum power that can be transferred from source to load.

Lastly, let us verify that given the load parameters of our **Thevenin circuit**, the original circuit of Figure 7.14, will deliver 58.8 watts to that load. We connect Z*Thevenin to the circuit in Figure 7.14, as shown in Figure 7.17.

Figure 7.17

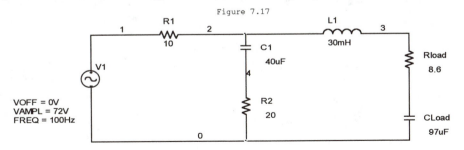

A **transient analysis** of 20 milliseconds produced the traces of V(1) and I(Rload) shown. The traces of V(3) and I(RLoad) are identical to those of the Thevenin equivalent circuit in Figure 7.16. It follows that the power delivered to the load in this case is identical to that delivered in the previous circuit.

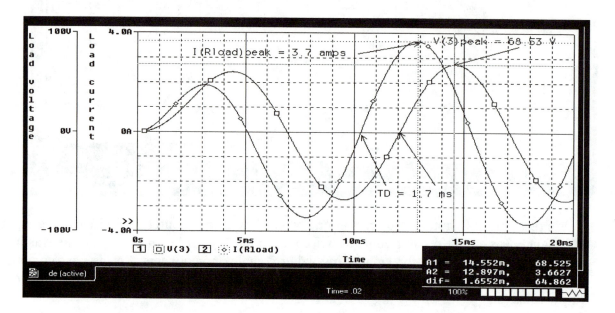

PROBLEMS

7.1   For the circuit shown, find the voltages across **RLoad** due to the sources **V1** and **V2.** Find the total voltage across **RLoad.** What power is contributed to **RLoad** by each voltage source? What is the total power delivered to **RLoad?**

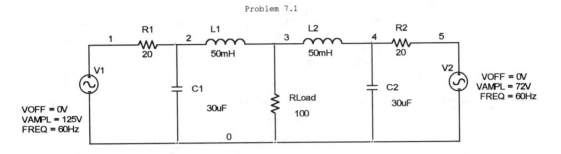

Problem 7.1

7.2   Repeat Problem 7.1 for the circuit shown.

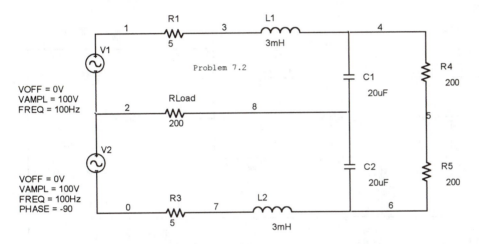

Problem 7.2

7.3   This circuit contains an independent voltage and an independent current source. Find the voltages contributed by either of these sources across **RLoad.** In addition, find the total voltage across **RLoad.**

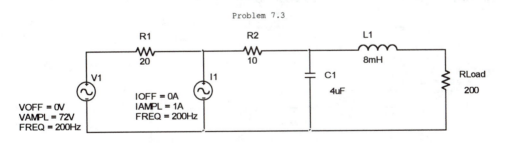

Problem 7.3

7.4   For this circuit, find the **Thevenin equivalent circuit** with respect to **RLoad**. Compare the results of the voltage obtained across **RLoad** from the Thevenin equivalent circuit with the voltage across **RLoad** obtained from the original circuit.

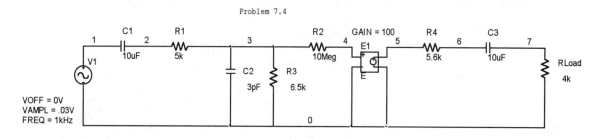

Problem 7.4

7.5   Find the voltage contribution that each voltage source makes to the load voltage across the complex load impedance. Also, find the total voltage across the load impedance.

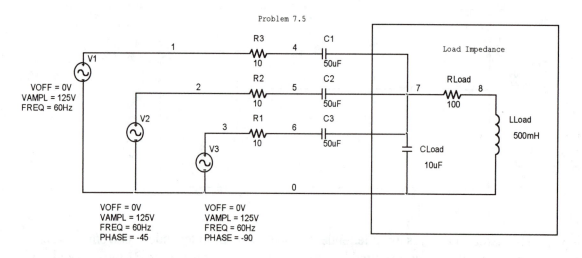

Problem 7.5

7.6   For this circuit, find the voltage across **RLoad** contributed by each voltage source. Also, find the total voltage across **RLoad**.

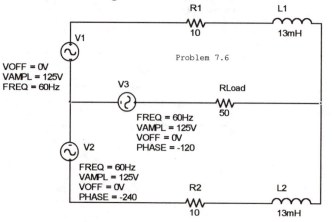

Problem 7.6

7.7   For this circuit, find the **Thevenin equivalent circuit** with respect to **RLoad**.

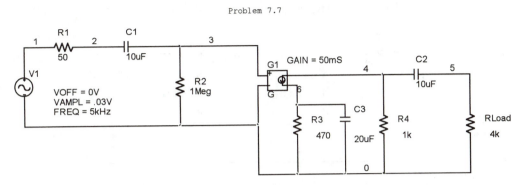

Problem 7.7

7.8   Use the same circuit as in Problem 7.7. The frequency of the voltage source **V1** has been changed to 5 Hz. Find the **Thevenin equivalent circuit** with respect to **RLoad.** In the previous problem, given the 5 kHz frequency of the voltage source, the effects of the capacitors in the circuit were essential that of short circuits. Thus, the equivalent circuit essentially was resistive. With the frequency changed to 5 Hz, this is no longer true.

7.9   For the bridge circuit in this problem, find the **Thevenin equivalent circuit** with respect to **RLoad.**

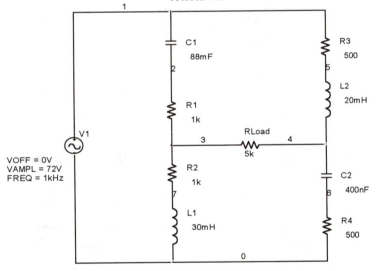

Problem 7.9

7.10   For this multisection circuit, find the **Thevenin impedance** with respect to **RLoad**.

Problem 7.10

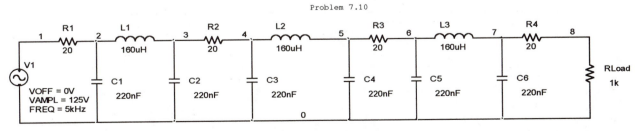

7.11   Obtain the **Thevenin equivalent circuit** with respect to **RLoad**. The circuit in Problem 7.11 contains a current-controlled voltage source of gain 20.

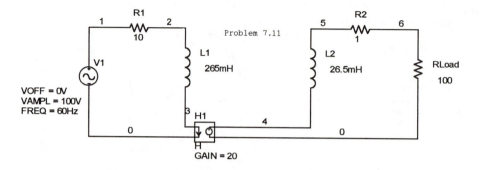

7.12   Find the **Thevenin equivalent circuit** with respect to the terminals of the complex load impedance. To obtain this equivalent circuit "by hand" can make one learn to appreciate the power of the **PSpice** program.

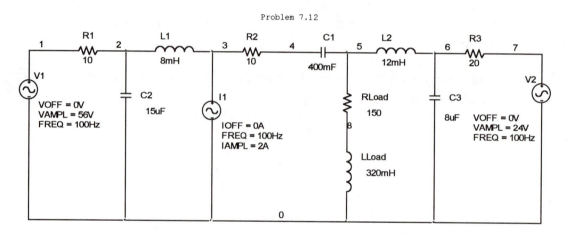

7.13   Find the Norton equivalent circuit with respect to the terminals of **RLoad**.

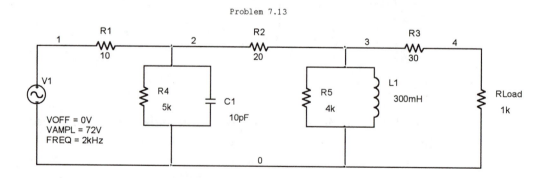

7.14   Find the Norton equivalent circuit with respect to **RLoad.**

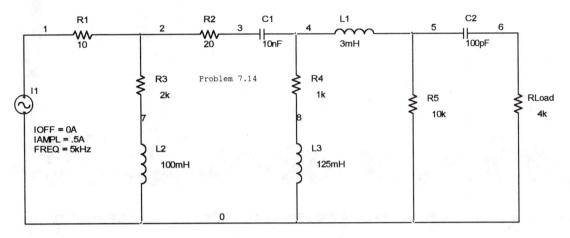

7.15   Find the **Norton equivalent** and the **Thevenin equivalent circuits** with respect to **RLoad.**

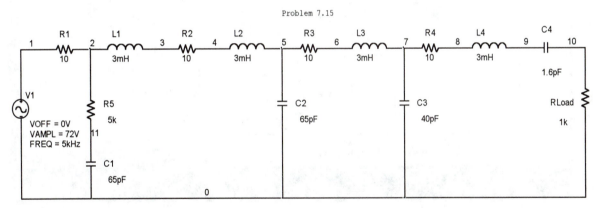

7.16   When **VTest** was connected to the circuit in this problem, the resulting **PROBE** traces of I(RTEST), the nodal voltage of V(4) of **VTest** and their RMS ratio are shown. From this data, find the needed circuit element to construct the **Thevenin equivalent circuit.**

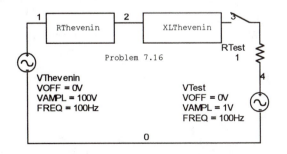

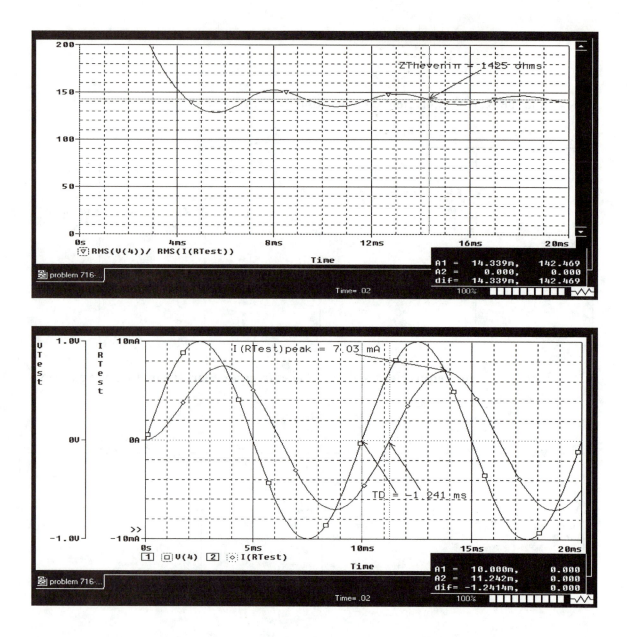

After the construction of the **Thevenin equivalent circuit** is completed, connect **VTest** in **series** with **RTest** and check out the design. **VThevenin** may be replaced by a short circuit. Make sure that the positive terminal of **RTes**t faces **VTest**. If this is not done, an unwarranted phase shift is introduced. between **VTest** and **RTest.**

7.17 Find the **Norton equivalent circuit** with respect to the terminals of **RLoad.**

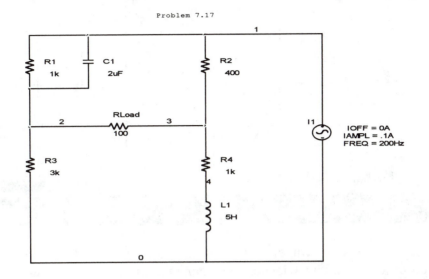

7.18 Find a load impedance that will permit maximum power to be transferred from
the voltage source **V1** to the load.

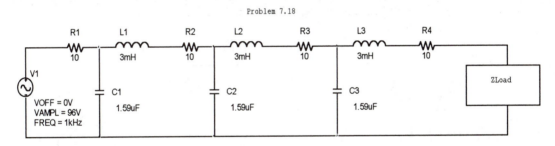

7.19 The Thevenin circuit shown delivers maximum power to the complex load
connected between nodes 3, 4 and zero.

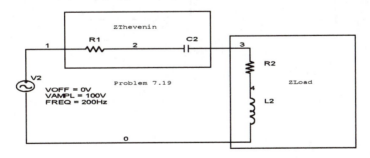

The **PROBE** data shown was obtained from this circuit. From it, determine the value of
the circuit elements needed for the condition of maximum power transfer.

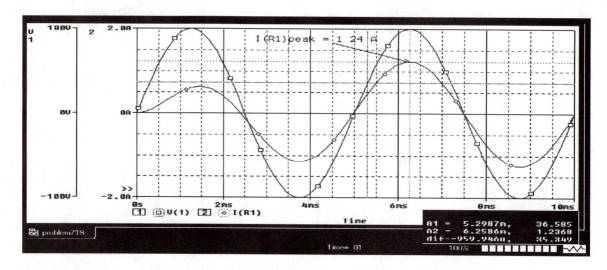

7.20 Find the load elements that will ensure maximum power transfer from the
      voltage source to the load impedance. Determine the value of that power.
      Explain the reason for the resistance value of **XLoad.**

Problem 7.20

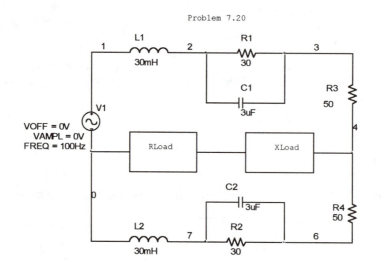

8

POWER AND ENERGY IN ALTERNATING CURRENT CIRCUITS

INTRODUCTION

In this chapter we shall investigate the power and energy relations in circuits containing capacitors, inductors and resistors. We apply sinusoidal current and/or voltage sources to these circuits. For a given load, the following expressions hold for its voltage across and current through it:

$$v(t) = V_p \sin \omega t \text{ volts}$$

$$i(t) = I_p \sin(\omega t + \theta) \text{ amps}$$

In these formulas, the subscript **p** denotes the peak amplitude of its associated quantity, ω is the radian frequency that we encountered before and θ is the impedance angle between the load voltage and load current.

The instantaneous power to the load is given by:

$$p(t) = v(t)i(t) \text{ volt-amperes}$$

By using various trigonometric identities, this product can be written as follows:

$$p(t) = V_p \frac{I_p}{2} \cos\theta (1 - \cos 2\omega t) + V_p \frac{I_p}{2} \sin\theta (\sin 2\omega t) \text{ volt-amperes}$$

Since $V_p I_p/2 = .707 V_P * .707 I_P$, we can write the above expression using the effective, or **RMS** values of the voltage and the current thus:

$$p(t) = V_{RMS}I_{RMS}\cos\theta(1-\cos 2\omega t) + V_{RMS}I_{RMS}\sin\theta(\sin 2\omega t) \text{ volt-amperes}$$

By convention, for an inductive load, the **impedance angle** θ is positive; for a capacitive load, that angle is negative. For a purely resistive load, that angle is zero degrees.

In addition to finding the instantaneous power to a load, we shall find the time average power flow to complex loads consisting of resistive and reactive elements. We shall introduce and use the concepts of real, reactive and apparent power and express their relationships using **PROBE**. In accord with traditional methods, we shall use the method of the power triangle to obtain the power flow in a circuit and use that method as a check against the solutions obtained from **PSpice** and **PROBE**.

POWER AND ENERGY FOR SINGLE PASSIVE ELEMENTS

Power and Energy for a Resistor

Let us obtain the power flow and the energy delivered over time to the resistor shown in the circuit of Figure 8.01.

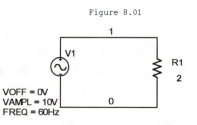

Figure 8.01

A **transient analysis** of 34 milliseconds produced the traces of the load voltage $V(1)$ and the load current $I(R1)$ shown. The peak current is 5 amps. No time delay or phase shift exists between $V(1)$ and $I(R1)$. The cosine of the angle θ between the load voltage and the load current is defined as the **power factor** of the load. For a purely resistive load, the angle θ is always zero degrees. This makes the **power factor** equal to unity. Physically it means that all power to the resistor will be converted into heat, none of it will be stored over time. Entering the phase angle of $0°$ into our expression for the instantaneous power shown above, we obtain

$$p(t) = v(t)*i(t) = V_{RMS}I_{RMS} - V_{RMS}I_{RMS}\cos 2\omega t \ \ watts$$

The power flow to **R1** consists of an average value of $V_{RMS}I_{RMS}$ watts minus a sinusoidal component that is equal in amplitude to the average value but oscillates at twice the frequency of the applied voltage source **V1**. Let us obtain a trace of that instantaneous power. It is shown below. The instantaneous power oscillates from a minimum of zero watts to a maximum of 50 watts at twice the applied frequency. It can be seen that its average is 25 watts. This quantity is equal to the first term in the above expression for p(t).

The instantaneous power never goes negative. This means that the resistor can never deliver, only receive power. Such is expressed by a **power factor** of unity.

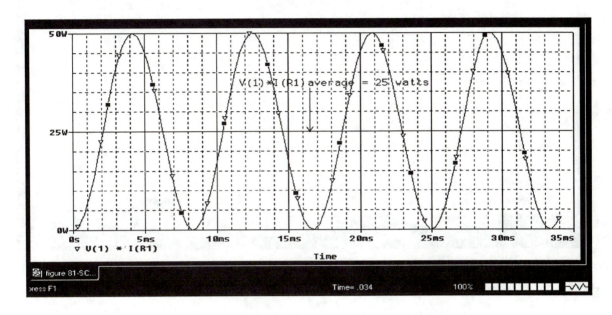

The energy W_e converted into heat by **R1** is equal to the integral of the instantaneous power over time. Symbolically:

$$W_e = \int p(t)dt \text{ joules}$$

As stated in Chapter 4, to obtain that integral, click on **Trace** and click on **Add Trace**. In the **Functions or Macros box**, click on **S( )**. It will appear in the **Trace Expression** box. Next, select variables V(1) and I(R1) to obtain the completed expression for the desired trace, as shown on the next **PROBE** plot.

The energy W_e, as shown below, increases linearly. This is due to the constant term $V_{RMS}I_{RMS}$, which we recall was equal to 25 watts. If we only consider the energy consumed over 34 milliseconds due to that term, we obtain:

25 watts* 34 milliseconds = .85 joules

This value corresponds closely to that indicated by its **PROBE** trace shown. This is the energy dissipated by the resistor during the 34 millisecond duration of our **transient analysis.**

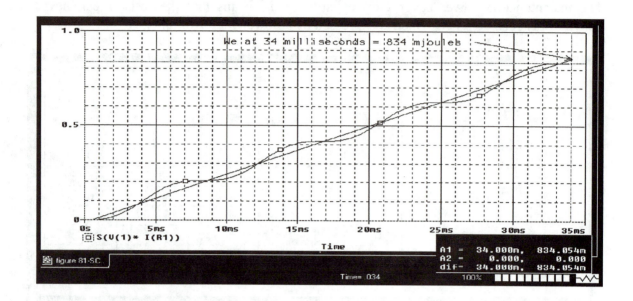

Power and Energy for a Capacitor

We shall next analyze the circuit in Figure 8.02.

Figure 8.02

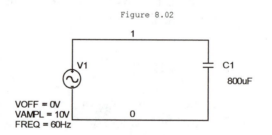

As we did for the resistor, we shall attempt to obtain the power flow and the energy stored in the capacitor over two cycles of the source voltage **V1**. A **transient analysis** of 34 milliseconds produced the traces of the capacitor voltage V(1) and it current I(C1). The current I(C1) has a peak value of 3.02 amps. The reactance of the capacitor C1 at 60 Hz is equal to

$$X_C = 1/(2*\pi*60*800 \text{ uF}) = 3.3 \ \Omega$$

If we divide the peak voltage of V(1) by the peak current of I(C1), we shall obtain the value of that reactance. To check: $10/3.02 = 3.3 \ \Omega$.

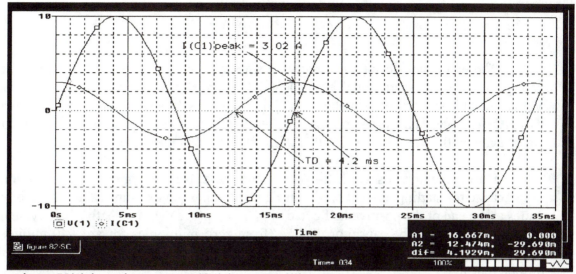

The voltage V(1) lags the current I(C1) by 4.2 milliseconds. This corresponds to a -90 ° phase angle of V(1) versus I(C1). Inserting this phase angle into our formula for p(t), and observing that sin(-90°) is equal to –1, we obtain for the instantaneous power p(t) :

$$p(t) = v(t)*i(t) = -V_{RMS}I_{RMS}\sin 2\omega t \quad volt\text{-}ampere \; reactive$$

Let us get a trace of this expression and that of the electrical energy W_e for the capacitor. The latter is given by

$$W_e = \int (-V_{RMS}I_{RMS}\sin 2wt)\,dt \quad joules$$

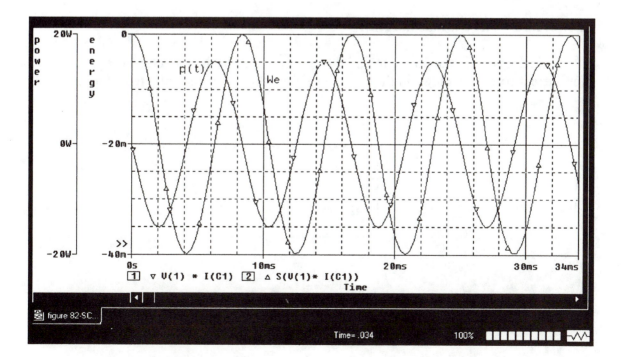

The expression for p(*t*) is a negative sine wave with a frequency twice that of the voltage source **V1** or its nodal equivalent V(1). Its amplitude is equal to the product $V_{RMS}I_{RMS}$. It is numerically equal to (.707)(10)(.707)(3.02) = 15.1 watts. This is shown on its **PROBE** trace. The expression for W_e, the integral of a sine wave, is a negative cosine wave, again with a frequency twice that of the source voltage **V1**.

During the first positive half-cycle of the power p(*t*), energy storage is building up in the capacitor starting at – 40 millijoules and attaining a maximum value of zero milli-joules at the end of the positive half cycle of p(*t*). As p(*t*) enters it negative half cycle next, the stored energy in the capacitor declines from zero millijoules toward –40 millijoules. That level of energy is reached at the end of the first negative half-cycle of p(t). Thereafter, events repeat.

The average value of the power flow is zero volt amperes as can be seen from its **PROBE** trace. We noted that the phase shift between capacitor voltage and current is -90°. This makes the **power factor** equal to zero. But a zero power flow over time means that the average energy of the capacitor cannot change. The trace of the energy of the capacitor has an average value of -20 millijoules over time.

To obtain their **PROBE** traces, we need only to enter the expressions for the product of the voltage and the current and the integral of that product to obtain the correct power flow and energy content for the capacitor. The value of the cosine and sine functions, given the impedance angle, are automatically calculated by the **PSpice** program.

Power and Energy for an Inductor

Let us obtain the power to and the energy stored in the inductor over two cycles of the current source **I1** shown in the circuit of Figure 8.03 by means of a **transient analysis**.

Figure 8.03

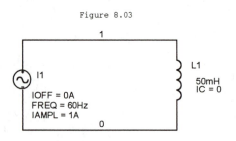

The resultant voltage V(1) has an amplitude of 18.9 volts. At the source frequency of 60 Hz, the reactance of **L1** is equal to

$$X_L = 2*\pi*60*50 \text{ mH} = 18.84 \text{ ohms}$$

This is equal to the amplitude ratio of V(1)/I1 = 18.9/1 = 18.9 ohms. The voltage V(1) leads the source current **I1** by a **TD** of 4.17 milliseconds, which corresponds to a phase lead of 90°. This is shown on the **PROBE** below.

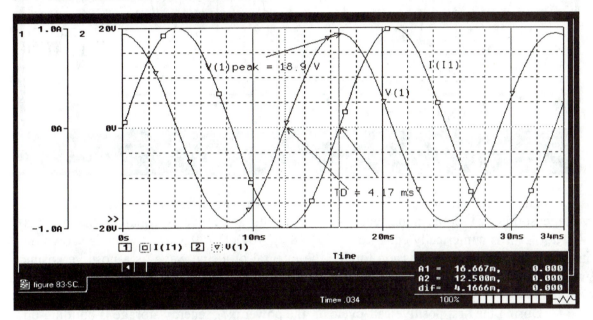

Since the cosine of an angle of 90° is equal to zero and the sine of an angle of 90° is unity, it follows that the expression p(t) for the power flow to the inductor reduces itself to

$$p(t) = V_{RMS}I_{RMS}\sin 2\omega t \quad volt\text{-}amperes \ reactive$$

This represents a positive sine wave with a frequency of 60 Hz. The magnetic energy W_m, supplied to the inductor over the two cycles of the current source **I1,** is equal to

$$W_m = \int (V_{RMS}I_{RMS} \sin 2\omega t)dt \quad joules$$

A **transient analysis** of 34 milliseconds produced the traces of p(*t*) and W_m shown below.

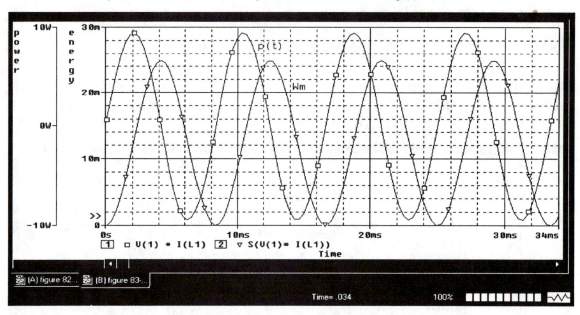

The expression for p(*t*) is a positive sine wave with a frequency twice that of the voltage source **V1** or its nodal equivalent V(1). Its amplitude is equal to the product $V_{RMS}I_{RMS}$. It is numerically equal to (.707)(18.9)(.707)(1) = 9.5 watts. This is shown on its **PROBE** trace. The expression for W_e, the integral of a sine wave, is a positive cosine wave, again with a frequency twice that of the source voltage **V1**.

During the first positive half-cycle of the power p(*t*), energy storage is building up in the capacitor, starting at zero millijoules and attaining a maximum value of 25 millijoules at the end of the positive half-cycle of p(*t*). As p(*t*) enters it negative half cycle next, the stored energy in the inductor declines from 25 millijoules toward zero millijoules. That level of energy is reached at the end of the first negative half cycle of p(*t*). Thereafter, events repeat.

The average value of the power flow is zero volt amperes as can be seen from its **PROBE** trace. The phase shift between inductor voltage and current is 90°. This makes the **power factor** equal to zero. But a zero power flow over time means that the average energy of the inductor cannot change. The trace of the energy of the inductor has an average value of 12.5 millijoules over time.

POWER AND ENERGY TO CIRCUITS

Power and Energy in an RC Circuit

We shall use the circuit in Figure 8.04 to find the voltage across and the current through the complex load. Also, we shall determine the phase angle of the load, the instantaneous and average power, and the energy storage in the capacitor. We are not done yet: we shall also introduce the concepts of real power, reactive power and apparent power.

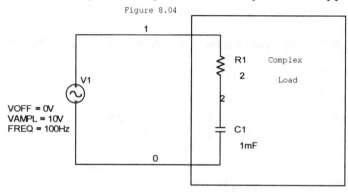

Figure 8.04

A **transient analysis** of 20 milliseconds produces the traces of the source voltage V(1), the resistor voltage V(1,2), the capacitor voltage V(2) and the circuit current shown as I(R1).

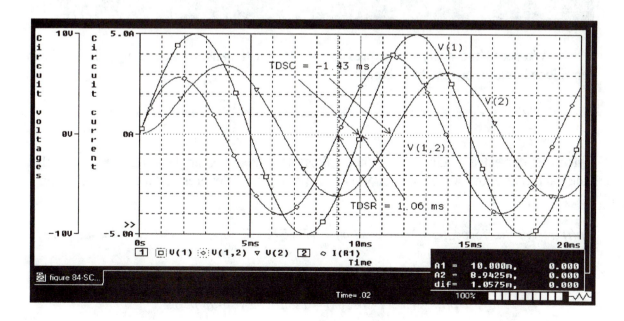

The resistor voltage V(1,2) and the circuit current I(R1) lead the source voltage V(1) by the indicated **TDSR** (**S**ource versus **R**esistor voltage) of 1.06 milliseconds. This corresponds to a phase lead of 38.2°. The capacitor voltage V(2) lags the source voltage V(1) by the indicated **TDSC** (**S**ource versus **C**apacitor voltage) of 1.43 milliseconds. This corresponds to a phase lag of 51.5°. The amplitude of the resistor voltage V(1,2) is 7.8 volts and the capacitor voltage V(2) is 6.2 volts. The current I(R1) is 3.9 amps. We shall use this data shortly to perform a **Phasor analysis** of this circuit. For now, let us obtain the power and the energy to the two circuit elements. The trace of the instantaneous power and the energy to the load is shown next.

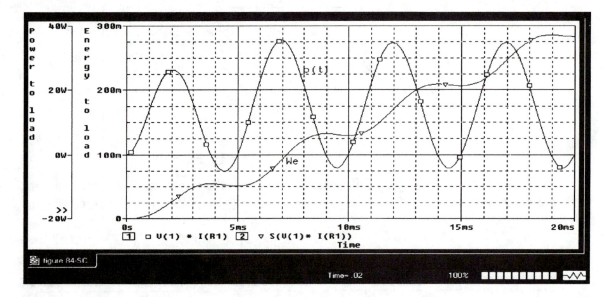

We recall the analytical expression for the instantaneous power p(*t*). It has been reproduced below for ease of reference.

$$p(t) = V_{RMS}I_{RMS}\cos\theta(1-\cos 2\omega t) + V_{RMS}I_{RMS}\sin\theta(\sin 2\omega t)\ volt\text{-}amperes$$

Since our current is leading V(1), this makes the impedance angle −38.2°. Entering that value and that of the RMS value of V(1) and I(R1) into the expression for p(t), we obtain:

$$p(t) = (7.07)(2.75)\cos(-38.2°)(1-\cos 2\omega t) + (7.07)(2.75)\sin(-38.2°)(\sin 2\omega t)\ volt\text{-}amperes$$

This simplifies to:

$$p(t) = 15.3(1-\cos 2\omega t) - 12\sin 2\omega t\ volt\text{-}amperes$$

The energy to the load is proportional to the positive area under the trace of p(t). Energy from the load back to the source is proportional to the negative area under the trace of p(*t*). Viewing our trace, we observe that the positive area is much larger than the negative one. That means that net energy is delivered to the load per cycle of **V1.** Thus, the energy increases over time. However we observe that during the time that the trace of p(t) goes negative, the slope of the energy trace dips negative. Thus it becomes clear that if such did not happen, the amount of the energy to the load would be increased.

This is a desirable condition. The energy returned to the source is a loss. That energy is not available to do useful work. Furthermore, electrical losses are incurred by the transmission of the energy back to the source. Utility companies do their best to achieve a power factor as close to unity as possible in order to keep these losses to a minimum. Compare the present trace of p(t) with that for the resistive load. In that case no area of the trace of p(t) was negative; its power factor was unity.

Time Average Power

Time Average Apparent Power to the Circuit: **S**

The product of the effective, or **RMS,** values of the load current times the load voltage is defined as the **apparent power** to the load. It is given the symbol **S** and its unit is the VA (volt-ampere). Depending on the amplitudes of the variables, the kVA (kilovolt-ampere) is used. For the circuit in Figure 8.04:

$$S = V_{RMS}I_{RMS} = (.707)(10)(.707)(3.9) = 19.5 \ VA$$

Time Average Real Power to the Circuit: **P**

The product of the effective, or **RMS** values of the load current time the load voltage times the cosine of the impedance angle is defined as the **real power** to the load. Its symbol is **P** and its unit is the W (watt). Depending on the size of the variables, the kW (kilowatt) is used. For the circuit in Figure 8.04:

$$P = V_{RMS}I_{RMS}cos(-38.2°) = 15.3 \ W$$

Time Average Reactive Power to the Circuit: **Q**

The product of the effective, or **RMS,** values of the load current times the load voltage times the sine of the impedance angle is defined as the **reactive power** to the load. Its symbol is **Q** and its unit is the VAR (volt-ampere-reactive). Depending on the size of the variables, the kVAR (kilovolt-ampere-reactive) is used.

For the circuit in Figure 8.04:

$$Q = V_{RMS}I_{RMS}\sin(-38.2°) = -12\ VAR$$

The Power Triangle

The relationship between the apparent, the real and the reactive time average powers can be expressed graphically by the power triangle. Since the real power and the reactive power are proportional to the cosine and the sine of the impedance angle respectively, they will be plotted at right angles with respect to each other as shown. The phasor sum of **P** plus **Q** will be equal to the apparent power **S**. The angle between **P** and **S** is the impedance angle of the circuit.

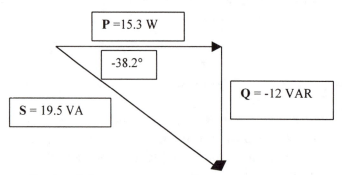

We can verify the amplitude of the apparent power:

$$S = \sqrt{P^2 + Q^2} = \sqrt{15.3^2 + (-12)^2} = 19.5VA$$

We can easily obtain the **PROBE** traces of **S, P** and **Q** as shown:

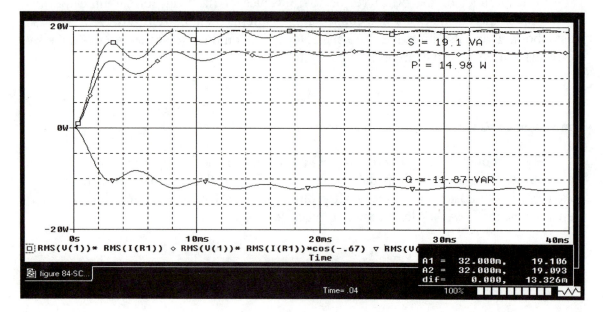

A **transient analysis** of 40 milliseconds was performed to allow V(1) and I(R1) to settle into their steady-state values. At about 32 milliseconds, the traces of **S**, **P** and **Q** have approximately attained their steady-state values. These are very close to those calculated above. To obtain these traces, the impedance angle had to be converted from -38.2° to -.67 radians.

It is instructive to rewrite the equation for p(t) in terms of the parameters **S**, **P** and **Q**. We restate the original equation:

$$p(t) = V_{RMS}I_{RMS}cos\ \theta(1\text{-}cos\ 2\omega t) + V_{RMS}I_{RMS}sin\ \theta(sin\ 2\omega t)\ volt\text{-}amperes$$

From the definition of these parameters, we can write two alternative forms of that equation:

$$p(t) = P(1\text{-}cos\ 2\omega t) - Qsin\ 2\omega t\ volt\text{-}amperes$$

alternatively,

$$p(t) = Scos\theta(1\text{-}cos2\omega t) + Ssin\theta(sin2\omega t)\ volt\text{-}amperes$$

This corresponds to the expression for *p(t)* obtained above.

Power and Energy in an RLC Circuit

We extend our analysis to include a circuit that has a resistor, a capacitor and an inductor. Although the circuit of Figure 8.05 is a series circuit, we recognize that this circuit can represent many circuits containing these three elements in any arbitrary fashion. Thus, the analysis of this circuit by extension covers these circuits also.

Figure 8.05

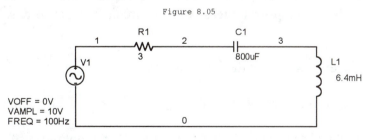

To start our analysis, the **PROBE** traces of the source voltage **V1**, or its nodal equivalent V(1), and the circuit current I(R1) are shown next. A **Transient analysis** of 40

milliseconds duration was used to allow both V(1) and I(R1) to settle into their steady state values.

Since the circuit contains both capacitive and inductive reactance, it is not obvious if the circuit current I(R1) leads or lags behind the voltage V(1). The phase between them is determined by the relative magnitudes of the capacitive and the inductive reactances at the source frequency of 100 Hz. At 100 Hz, the capacitive reactance is 2 Ω and the inductive reactance is 4 Ω. Thus, the latter determines the sign of the phase angle. The traces of V(1) and I(R1) are shown next.

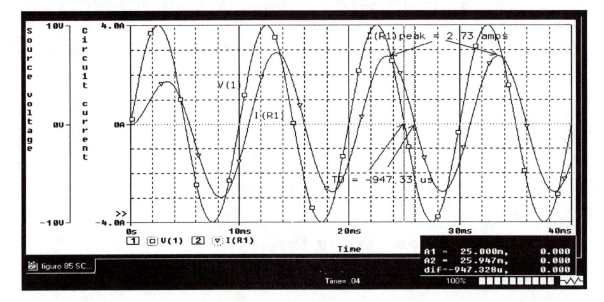

The peak amplitude of the current I(R1) is equal to 2.73 amps. The current lags the voltage V(1) by 947.33 microseconds. This corresponds to a phase lead of 34.1°. The lagging current, which gives rise to a positive impedance angle, verifies the point that the inductive reactance dominates over the capacitive reactance. This will be brought out by the power triangle also. The **power factor** of this circuit is equal to cos(34.1°), which is equal to .83.

We recall the expression for the instantaneous power of p(t):

$$p(t) = V_{RMS}I_{RMS}\cos\theta(1-\cos 2\omega t) + V_{RMS}I_{RMS}\sin\theta(\sin 2\omega t) \text{ volt-amperes}$$

If we enter the phase angle of 34.1° and the **RMS** amplitudes of V(1) and I(R1), we obtain:

$$p(t) = (.707)(10)(.707)(2.73)(.83)(1-\cos 2\omega t) + (.707)(10)(.707)(2.73)(.56)(\sin 2\omega t)$$

$$p(t) = 11.4(1-\cos 2\omega t) + 7.6\omega t \ \text{volt-amperes}$$

We recall from our previous circuit that the coefficient of the first term in this equation is equal to the real power, **P.** It is equal to 11.4 watts. The coefficient of the second term is equal to the reactive power **Q**. It is equal to 7.6 volt-ampere-reactive.

We shall next plot $p(t)$. We need only enter the product V(1)*I(R1) to obtain the trace of this expression. The **PROBE** program will take care of the amplitudes of the variables and their respective phase angle. In addition, we shall get the trace of the average value of $p(t)$ and the energy flow for the circuit for the duration of the 40 millisecond **Transient analysis**. All of this is shown next.

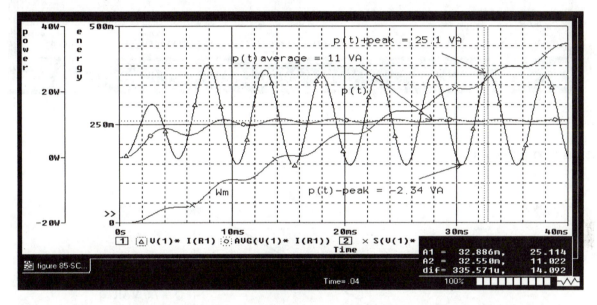

The instantaneous power has a peak-to-peak amplitude of 27.44 volt-amperes or a peak amplitude of 13.72 volt-amperes. Its average value is equal to 11 volt-amperes. The fact that the negative peaks of $p(t)$ dip below the zero value on the power axis means that its average is reduced from 13.72 volt-amperes to 11 volt-amperes. Incidentally, the average power is the power delivered to the resistor. The energy delivered to the circuit increases linearly on average. As in the previous example, for every instance that $p(t)$ is negative, the slope of the energy trace goes negative.

We next determine **S**, **P** and **Q** using **PROBE**. Their traces are shown next. Their values were obtained at 30 milliseconds when steady-state was reached. The apparent power **S** is 13.43 VA, the real power **P** is 11.3 watts, and the reactive power **Q** is 7.6 volt-ampere. If we compare the values of **P** and **Q** with the coefficients of $p(t)$ above, we see that they are the same.

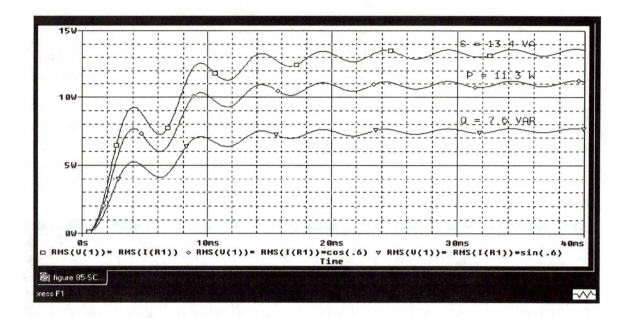

Let us next plot the power triangle for this circuit.

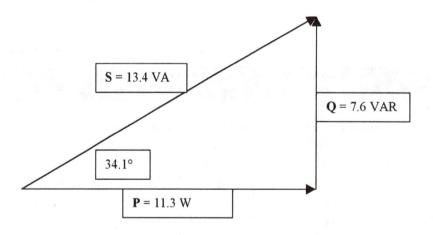

The positive phase angle causes the reactance to be drawn upward as shown. This is indicative of the fact that the net reactance in this circuit at the source frequency of 100 Hz is inductive. Such may not be the case if the source frequency is changed. That condition will occupy us in a future chapter. If we compare this power triangle with that of the previous circuit, we see that in the former case, the net reactance was capacitive and it was the only kind in the circuit: hence, its **Q** was drawn at right angles to **P** in a downward direction.

We finally verify the amplitude of the apparent power S using the data from the power triangle.

$$S = \sqrt{P^2 + Q^2} = \sqrt{11.3^2 + 7.6^2} = 13.4 VA$$

This result shows good agreement with the PROBE data.

Power Factor Correction

We have previously emphasized the fact that it was desirable in the transfer of energy to a load to have a **power factor** as nearly to unity as possible. That would entail that the apparent power **S** is close to the real power **P**. Let us study the power and energy of the circuit shown in Figure 8.06. As will be shown, it has a **power factor** considerably less than unity. Our effort is directed to result to achieve a power factor of .9. We begin by analyzing the existing power and the **power factor** of this circuit.

Figure 8.06

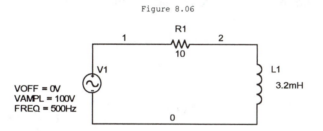

A **transient analysis** of 4 milliseconds produced the traces of V(1) and I(R1).

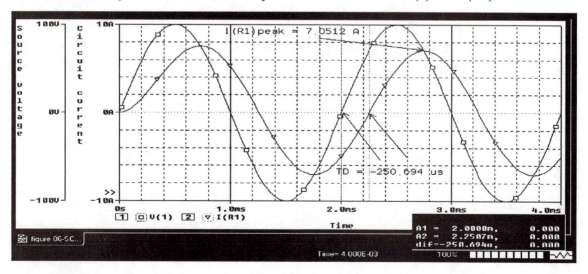

Not surprisingly, the current I(R1) lags the voltage V(1) by 250.694 microseconds. This corresponds to an impedance angle of 45.1°. Its associated **power factor**, cos(45.1°), is equal to .705.

Let us improve this power factor to the desired value of .9. At 500 Hz, the inductive reactance $X_L = 10\ \Omega$. To obtain a power factor of .9, we need an impedance angle of arccos(.9) = 25.84°.

To obtain the needed reactance at that angle, defined as X_T, We evaluate

$$\tan(25.84°) = X_T/R = .48 = X_T/10$$

From which follows

$$X_T = (.48)*10 = 4.8\ \Omega$$

To reduce the original reactance of the circuit, determined by $X_L = 10\ \Omega$ to the needed level, we need to add capacitive reactance X_C by adding a capacitor to the circuit. It remains to determine the needed size of that capacitor. Since

$$X_T = X_L - X_C = 4.8\ \Omega$$

It follows that

$$XC = XL - XT = (10 - 4.8)\ \Omega = 5.2\ \Omega$$

Finally

$$C = 1/(2*\pi*f*X_C) = 61.2\ uF$$

Let us enter a capacitor of this capacitance into our circuit, shown in Figure 8.07, and obtain the new **power factor.**

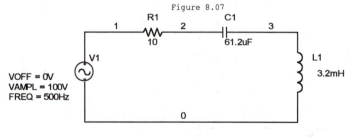

Again, we run a **transient analysis** of 4 milliseconds and obtain the traces of V(1) and I(R1).

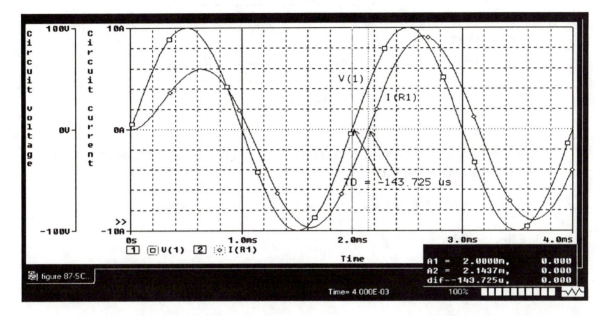

The new **TD** shows that the current I(R1) is lagging behind V(1) by 143.725 microseconds. This corresponds to a phase lag of 25.87°. The associated **power factor** is equal to cos(25.87°). The cosine of that angle is .9. Our objective has been achieved.

Power and Energy in a Circuit with A Transformer

What's new?

1. The transformer in **PSpice**

A transformer is a four-terminal device containing two circuits that are magnetically coupled. Among their uses are the change of voltage levels within often quite complex systems, such as the electrical power grid of a city. A more familiar application at the consumer level is the myriad of transformers that are used to connect various low-voltage applications to the house current. Another application, one that we shall study, is to match the impedance of a load to a source to insure maximum power transfer between them.

The circuit in Figure 8.08 has a transformer with a primary voltage of 120 V and a secondary voltage of 60 V, which appears across **Rload**. Such a transformer is defined as a step-down transformer. We shall analyze this circuit to obtain the power delivered by the voltage source V1 to **Rload.**

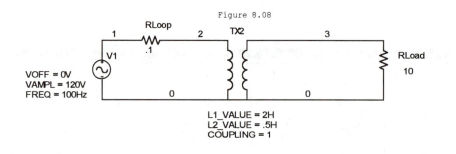

Figure 8.08

1. To obtain the transformer, in **Place Part** dialog box, click on **ANALOG** in **Libraries** box to select it.

2. Scroll to **XFRM LINEAR;** click on the symbol to select it.

3. Click on **OK**.

4. Place the part in the usual manner at the desired location on **Schematic screen**.

We must next set the parameters **COUPLING, L1_VALUE AND L2_VALUE.** They are defined as follows: **COUPLING** is the coefficient of magnetic coupling between the primary and the secondary winding of the transformer. A typical value of .99 means that most of the magnetic flux from the primary winding links the secondary one. We shall assume a perfect linking by setting that parameter equal to unity.

The Parameters L1_VALUE and L2_VALUE define the inductances of the primary and secondary windings, respectively. What is critical in their selection is not their absolute value but their ratio. We start with the fact that the secondary voltage of 60 volts is one-half the primary voltage of 120 volts. From this follows that the turns ratio of the transformer is equal to 2. The inductance of a coil is proportional to the square of the number of its turns. From this it follows that the inductance ratio is equal to the square of the turns ratio, that being equal to 4. Let us set the value of L1_VALUE equal to 2 H; hence, L2_VALUE is set at .5 H. These values are set as follows:

1. Double click on the symbol of **XFRM LINEAR. The ORCAD Capture-[Property Editor]** dialog box will appear on the **Schematic** screen.

2. In **COUPLING** box, enter 1, scroll to **L1_VALUE**, enter 2H, scroll to **L2_VALUE**, enter .5H.

3. Click on **Display…** . In the **Display Properties** box, **click on Name and Value**, click on **OK.**

4. Click on **Apply.**

5. Close the **Properties Editor.** The specified parameter values will appear on the **Schematic** screen.

We are now ready to run a 20 millisecond **transient analysis**. The **PROBE** traces of the primary voltage V(1), the primary current I(RLoop), the secondary voltage V(3) and the secondary current I(RLoad) are shown next.

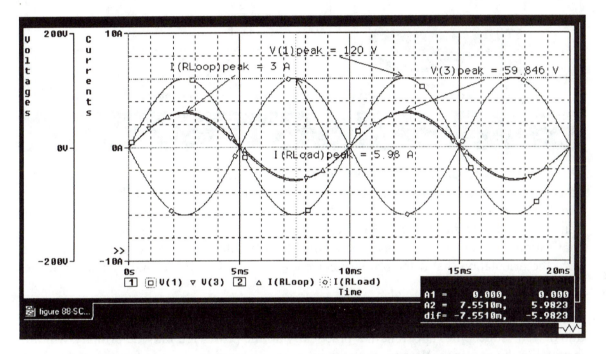

The secondary voltage V(3) at 59.6 volts is near its theoretical value of 60 volts. The secondary current at 5.96 amps is near its theoretical value of 6 amps. The primary voltage is at 120 volts and the primary current is at 3 amps. Since this is a step-down transformer, the secondary voltage is one-half that of the primary voltage, whereas the secondary current is twice that of the primary current. Also observe the 180° degree phase shift between the primary and the secondary current.

Let us next study the power and energy in this circuit. The power from the primary circuit and to the secondary circuit and their respective average values are shown next. Note on the plot that the cursor box has been moved. This was done because its location obstructed the expression for the last trace. To move it, simply place the cursor into the blue band on top of the cursor box and drag it to and drop it in desired location.

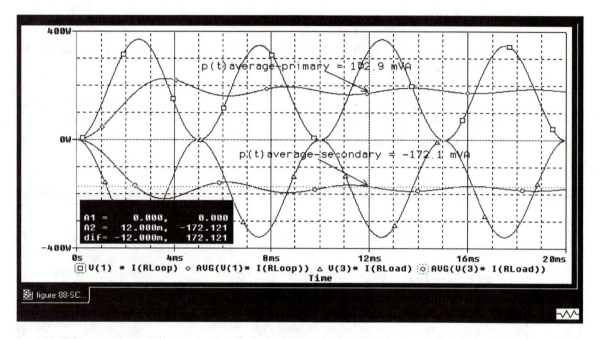

The power delivered by the voltage source into the primary winding of the transformer is equal in magnitude but of opposite sign to that delivered by the secondary winding to Reload. The reason that these two powers are of opposite sign is that the transformer model in **PSpice** has negative mutual inductance. In terms of the dot convention that is used in the analysis of transformers, the windings in the **PSpice** model have the orientation that if the primary current enters its dot, the secondary current leaves its dot.

Our concern here is not so much with the signs as with the amount of power transferred. It becomes apparent that practically all the power fed into the primary winding is equal to the power to **RLoad**. Since the power delivered is equal to the power received, we can anticipate that the energy delivered by the voltage source **V1** is equal to the energy dissipated by **RLoad.** This is confirmed by the traces shown next: the total energy delivered by the voltage source **V1** after 20 milliseconds of 3.59 joules is numerically equal to the energy dissipated during that same interval.

A final word on that transformer model. The term **LINEAR** denotes a transformer without a magnetic core. This assures linear operation since the saturation effect of an iron core is not present. However, the price to be paid is that of a lower **COUPLING** parameter. This means that not all flux lines link both the primary and the secondary windings. Values of the **COUPLING** parameter of about .4 are more typical.

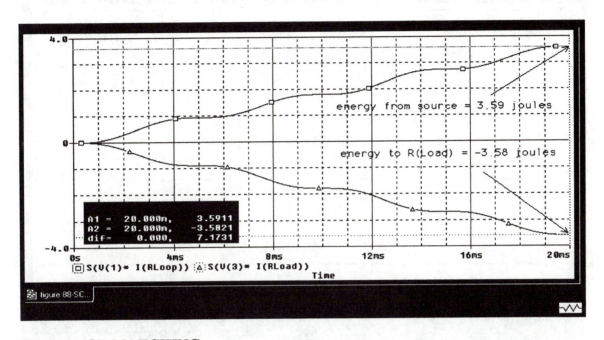

IMPEDANCE MATCHING

The circuit in Figure 8.09 consists of a lossy voltage source **V1** with an intrinsic **RSource** of 100 Ω and an **RLoad** of 10 Ω. It is desired to transfer maximum power from source to load. Neither of the two resistors can be changed. However, the condition for maximum power requires that **RLoad** is equal to **RSource**. We shall interpose a transformer between source and load to match the resistances of **RSource** and **RLoad.**

Figure 8.09

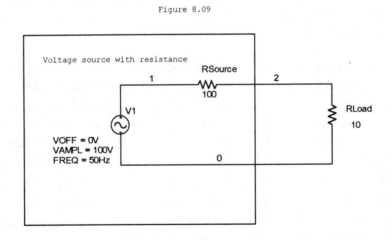

We shall first obtain the **RMS** power to **RLoad** using the existing circuit. This will allow for a comparsion of the improvement we hope to make by altering the circuit.

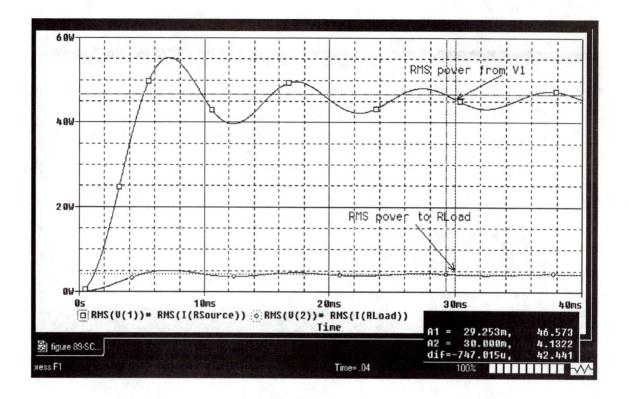

The RMS power from the voltage source **V1**, measured at 30 milliseconds, is about 47 watts, and the **RMS** power to **RLoad** is only about 4.1 watts. Thus, the ratio of power received to power delivered is 4.1/47, which is equal to .087, or only 8.7%. This is not an acceptable power transfer ratio. Let's do better.

The key to success is to make **RLoad** appear to the primary circuit with the same resistance as **RSource**. This would fulfill the conditions of maximum power transfer. Here is how we proceed. The resistance of **Rload** is reflected into the primary circuit as a^2**RLload**. In this formula, "a" is the transformation ratio, defined as N_P/N_S. N_P is the number of turns in the primary winding, and N_S is the number of turns in the secondary winding of the transformer. For optimum power transfer, the product a^2**RLoad** should be equal to 100 Ω, from which it follows that we need a step down tranformer with a transformation ratio equal to $\sqrt{10}$. It remains to determine the inductances of our transformer. We recall that the inductance of a coil is proportional to the square of the transformation ratio. In the symbols of our transformer

$$L1\_VALUE = a^2 L2\_VALUE$$

Setting L2_VALUE = 10 H

$$L2\_VALUE = 10L1\_VALUE = 10(10 \text{ H}) = 100 \text{ H}$$

We now modify the circuit in Figure 8.09 as shown in Figure 8.10.

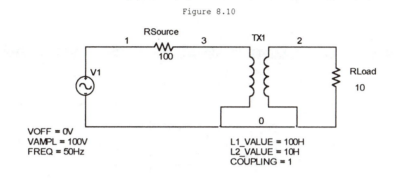

Figure 8.10

Observe the settings of the transformer parameters. We next run a **transient analysis** of 40 milliseconds to obtain the **RMS** powers of source and load.

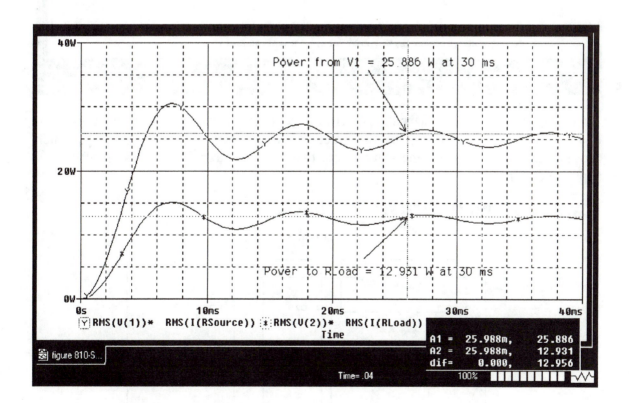

The trace for the power to **RLoad** shows an **RMS** value of 12.931 watts, whereas the trace of the RMS power from the voltage source V1 is at 25.988 watts. Thus it appears that about half the power of that source is dissipated in **Rload** and the other half in the resistor **RSource.** This is, of course, the condition of maximum power transfer from a resistive source to our load. This is the best power transfer we can hope for given the circuit in Figure 8.10. We have achieved an increase of power to **RLoad** equal to (12.9 – 4.1) watts = 8.8 watts. This corresponds to an increase in power of 215 %.

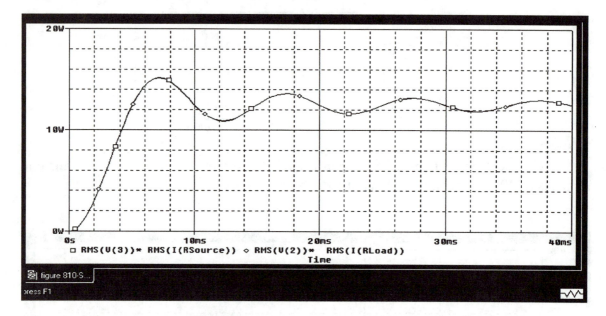

When we compare the actual power into the transformer compared to the power out of the transformer, we obtain the two traces shown. They completely overlap, which proves that they have the same value. Note that the ideal transformer has an efficiency approaching unity.

POWER IN A CIRUIT WITH SPECIFIED LOAD CONDITIONS

Figure 8.11

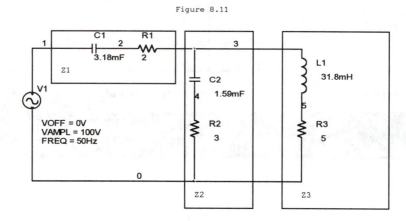

The circuit in Figure 8.11 has the real and reactive power to each circuit impedance listed next. The calculated apparent powers are also listed. The last row contains the sum of the quantities for each column. The question is: is it correct to add these powers as was done in their tabulation? We shall perform a **PSpice** analysis to find out.

Impedance	Real Power	Reactive Power	Apparent power
Z1	P1 = 300 W	Q1 = -150 VAR	S1 = 340 VA
Z2	P2 = 400 W	Q2 = -275 VAR	S2 = 500 VA
Z3	P3 = 80 W	Q3 = 150 VAR	S3 = 170 VA
Totals	P(total) = 780 W	Q(total) = -275 VAR	S(total) = 1010 VA

We start our analysis by obtaining the impedance angle of the voltage V(1) relative to the current I(R1). A **transient analysis** of 40 milliseconds produced their traces shown below. The voltage V(1) lags the current I(R1) by .968 milliseconds. This corresponds to a −17.4° impedance angle.

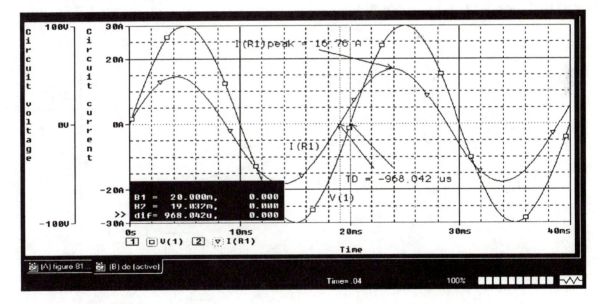

We next obtain the apparent power, the real power and the reactive power of this circuit using **PROBE**. To obtain their traces, we need to convert the −17.4° of the impedance angle to an angle of - .30 radians.

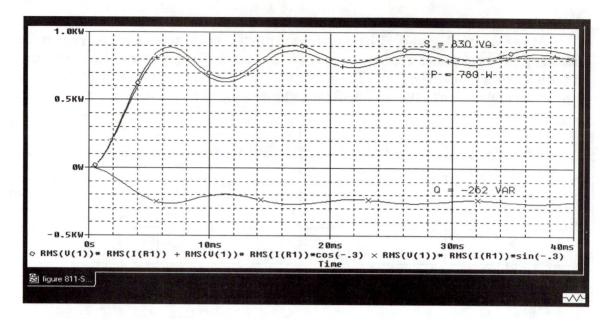

The traces of the total real power **P(total)** of 780 watts and the total reactive power **Q(total)** of -262 VAR correspond closely to those of our tabulation above. However, the apparent power **S(total)** of 830 VA differs substantially from its tabulated value. To resolve the discrepancy between the tabulated apparent power and its value obtained from **PROBE** data, we turn to the power triangle.

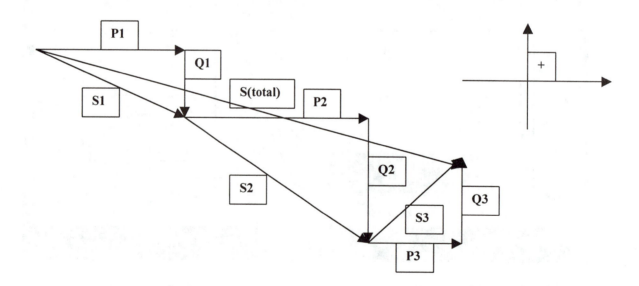

From this diagram it becomes apparent that all the real powers, **P1**, **P2** and **P3**, are parallel to each other and in the same direction. Therefore, to obtain the total real power, we need only add them. The reactive powers, **Q1**, **Q2** and **Q3,** are also parallel to each other; however, **Q1** and **Q2** are negative and **Q3** is positive. To obtain their sum, we must add them algebraically. Finally, the apparent powers, **S1**, **S2** and **S3,** are not parallel to each other. They must be added as phasors. Thus, the total apparent power, **S(total),** in our tabulation is wrong because it was obtained by simply adding the constituent apparent powers to the circuit impedances. There is one special case in which the apparent powers can be added: if the impedance angles of all impedances are the same. This would make all the apparent powers connect in a straight line.

As a final check, let us compute the apparent power **S(total)** into our circuit from the tabulated data and compare it with the value of its **PROBE** trace.

$$S(total) = \sqrt{780^2 + (-275)^2} = 827VA$$

This value compares closely with the 830 VA obtained from its **PROBE** trace.

PROBLEMS

8.1   For this circuit the time-average power delivered to each resistor by the two voltage sources. How much of this power is contributed by each source? Resistors RLine 1 and RLine 2 are the conductors that connect the voltage sources to the   load resistors. What is the percent efficiency of power transmission?

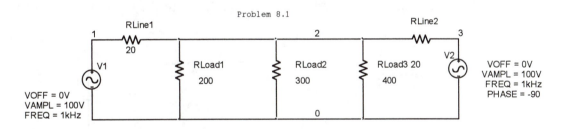

8.2   A sinusoidal voltage source with an internal resistance **RSource** of 50 Ω is supplying power to an **RLoad** of 2 kΩ. That resistor will convert all received energy into heat. Consequently, it may just get too hot! To ensure that **RLoad** does not burn out, we must compare the power rating of the resistor with its time average power received. Note that the peak power does not have sufficient time to heat the resistor to a dangerous degree; however, its **RMS** power may. The task at hand is to find the **RMS** power to **RLoad** and to specify the needed power rating for **RLoad.**

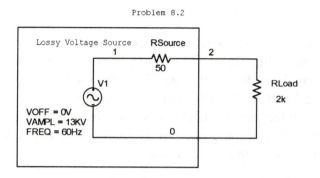

Problem 8.2

8.3   For this circuit, determine the instantaneous power to **RLoad.** Obtain the real, the reactive and the apparent power to **RLoad.** Get the power factor of this circuit. Draw the power triangle. What percentage of the real power is delivered to **RLoad**?

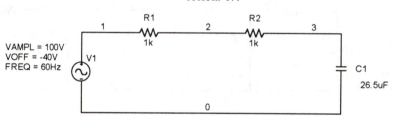

Problem 8.3

8.4   What is the input impedance of this circuit? Obtain the instantaneous and the **RMS** current of this circuit. Find the real powers to the two resistors. Get the reactive and the apparent power for this circuit. What are the impedance angle and the power factor of this circuit? Draw the power triangle. Note the nonzero offset voltage of **V1**.

Problem 8.4

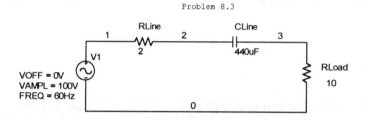

8.5   Obtain the instantaneous power and the energy flow of this circuit for two cycles of the input voltage **V1.** Find the real power to each resistor and the total real power to the circuit. Get the reactive power to each inductor. Find the percentage of total magnetic energy stored in each of the two inductors.

Problem 8.5

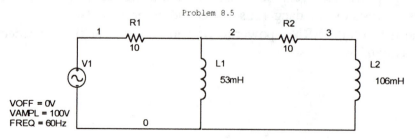

8.6   What are the load voltage and the load current in this circuit? Obtain the amplitude and the impedance angle of the input impedance. What is the **power factor** of the this circuit? Find the real and reactive power delivered to the load by **V1**. Find the real and the reactive power delivered to the load by **I1**. What is the sum of these powers? Of the total real power into the circuit, what percentage is converted into heat by the load?

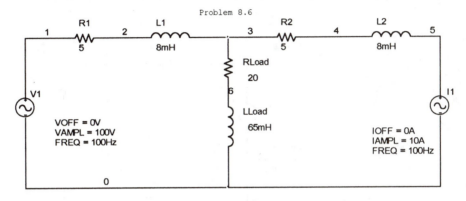

Problem 8.6

8.7   For the circuit shown, find the Thevenin equivalent circuits with respect to the complex load. Determine the total real power to the circuit and to **RLoad**. Determine the total reactive power to the circuit, **LLoad,** and to the voltage source, **V1**. What is the **power factor** of the circuit?

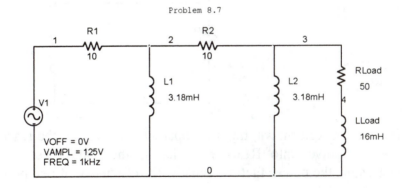

Problem 8.7

8.8   A 25 kV sinusoidal voltage source is applied to a circuit that absorbs 100 kW of real power and 300 kVAR of inductive reactive power. Construct the simplest series RL circuit that will account for this power. Check your design with a **PSpice** analysis.

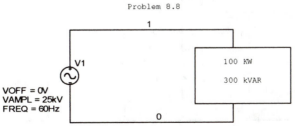

Problem 8.8

8.9   For this circuit, find the apparent, the real and the reactive powers into the circuit. What is the **power factor**? After the initial analysis is completed, add a 318 uF capacitor across the complex load. Repeat the analysis and tell what changes in the above powers and the **power factor** have occurred. Explain your findings.

Problem 8.9

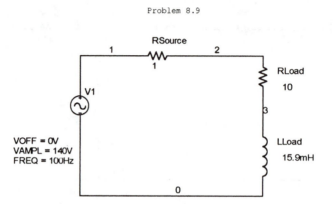

8.10   It is desired to improve the power factor of this circuit to .8 by adding a capacitor in series with the inductor **L1**. Find the value of that capacitor and check your analysis to make sure that the desired **power factor** has been achieved.

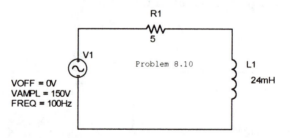

8.11   For the network shown, find the apparent, the real and the reactive power. What is the real power into **RLoad?** What is the real power lost in the network? Determine the **power factor**. What is the efficiency of real power transmission for this network?

Problem 8.11

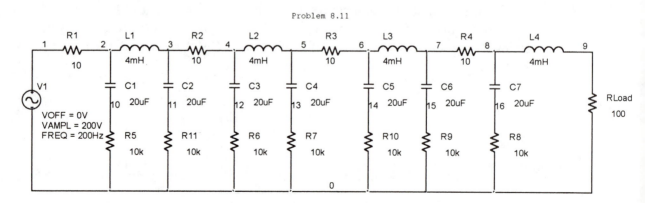

8.12   Two sinusoidal voltage generators each supply **RLoad** via two 1 Ω cables. When the instantaneous power from the two voltage sources to **RLoad** was plotted, the **PROBE** traces shown were obtained.

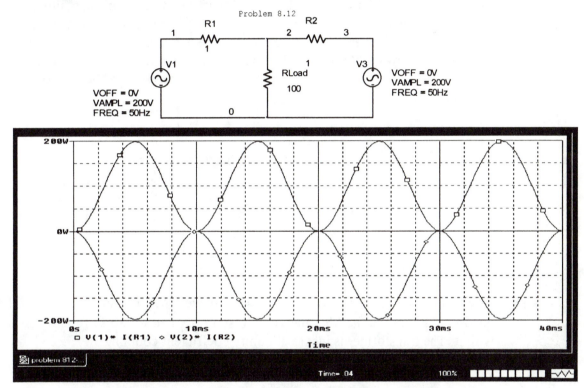

Why is the instantaneous power delivered by source **V2** the negative of the power delivered by **V1**? What way is there to make the power delivered by **V2** positive? Do it, and run the analysis again.   Obtain the sum of the instantaneous powers delivered by the two sources using the trace: V(1)*I(R1) + V(2)*I(R2).   You can simply type the plus sign or get it from the **Macros** box. Obtain the total **RMS** power delivered to **RLoad** and the energy over two cycles of the voltage sources. What is the percent efficiency of power transmission of this system?

8.13   In Problem 8.12, the **PHASE** of voltage source V2 has been changed to -30°. For this condition, answer the same questions as in Problem 8.12, In addition, compare the answers to each question in both problems and state what the differences in the answers are. In particular, which of the two voltage sources delivers most of the power to the circuit? Incidentally, changing the phase angle of the voltage sources relative to other voltage sources is a technique that utility companies use to direct the power flow within their systems. The name given to this technique is phase angle regulation.

8.14 The circuit in this problem is that of a simplified transistor amplifier. At the frequency of 1 kHz, the model is purely resistive. The circuit contains a current-controlled current source with a gain of 100. Find the **RMS** power to **RLoad** and the **RMS** power delivered by the voltage source **V1**. Can you explain the difference between these two **RMS** powers? That question is not entirely fair, but, give it a try! It is well to remember the reason for the difference in these two powers in your future work.

Problem 8.14

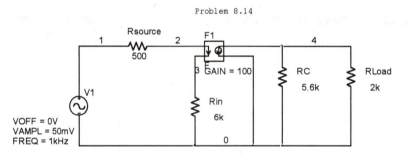

8.15 Design the transformer in this circuit so that the RMS power to **Rload** will be a maximum given the fact that the voltage source V1 has a 50 Ω resistor associated with it. The transformer is ideal with unity coefficient coupling.

Problem 8.15

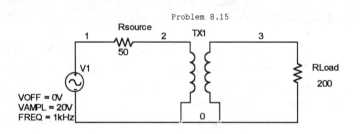

8.16 For this system, find the **RMS** power delivered to **RLoad**. Note: the transformation ratios for the two transformers are indicated. Assume that the transformers are ideal. You are free to choose the inductance levels. Find the efficiency of power transmission from the voltage source V1 to **RLoad.**

Problem 8.16

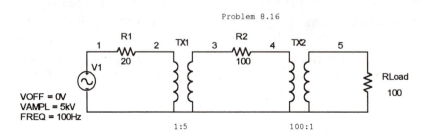

8.17 This circuit is at resonance. Find the apparent, the real and the reactive powers into the circuit. What is the total real power delivered by the source voltage and that absorbed by the circuit. Plot the instantaneous reactive powers of the inductor and the capacitor for two cycles of the input voltage on one **PROBE** plot. What is their relationship? Obtain the energy to the resistor, the inductor and the capacitor over two cycles of the input voltage. What is the input impedance of the circuit? What is the impedance angle and the **power factor** of the circuit?

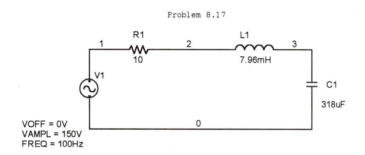

Problem 8.17

8.18 Connect the circuit elements from the previous problem as shown in this circuit. The voltage source has not changed. Find the input impedance of this circuit by running a **Transient analysis** for 200 milliseconds. Compare the source voltage with the voltage across the capacitor and the inductor. Explain your findings relative to the input impedance of this circuit. Find the instantaneous powers to both the capacitor and the inductor. Determine the **RMS** value of those powers. What does your finding imply? Finally, compare all voltages and currents in this circuit with those of the previous problem. Which are the same? Which have changed?

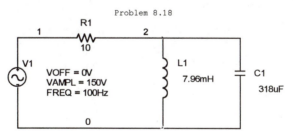

Problem 8.18

9

FREQUENCY RESPONSE OF RC, RL AND RLC CIRCUITS

TRANSIENT AND FREQUENCY RESPONSE OF AN RC CIRCUIT

The Transient Response

We shall introduce the topic of **frequency response**, in **PSpice** referred to as **AC analysis,** by first performing a **transient (time)** response of an **RC** circuit. Voltage source **V1** is connected to the circuit in Figure 9.01. Its amplitude is 10 V and its frequency is 100 Hz. We shall study the resultant **transient response** for two cycles of that frequency for each applied frequency. This will lead to the major topic of this chapter.

Figure 9.01

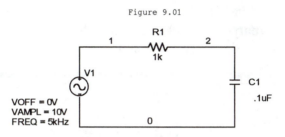

The result of a 20 millisecond **transient analysis** is shown next. The capacitor voltage V(2) is almost identical to the source voltage **V1** or its nodal equivalent V(1). They are plotted on the number 1 Y axis. The resistor voltage V(1,2) at its peak is only .63 volts. It is plotted on the number 2 Y axis. It leads the source voltage V1, or its nodal equivalent by a **TD** equal to 2.4 milliseconds. This corresponds to an 86.4° phase lead.

At a frequency of 100 Hz, the reactance of the capacitor is 15.9 kΩ. Hence the peak amplitude of V(2) is determined by the voltage divider ratio of 15.9 kΩ/16.9 kΩ which is equal to .94. Thus the peak amplitude of V(2) is equal to (.94)(10 V) = 9.4 V. The traces of V(1) and V(2) are almost identical. This means not only are their amplitudes almost the same, their relative phase shift is zero degrees.

We can intuit that if we decrease the frequency further, the reactance of the capacitor will approach infinity ohms and no current will flow in the circuit. Thus the resistor voltage V(1,2) will be zero volts and the capacitor voltage V(2) will be equal to the source voltage V(1).

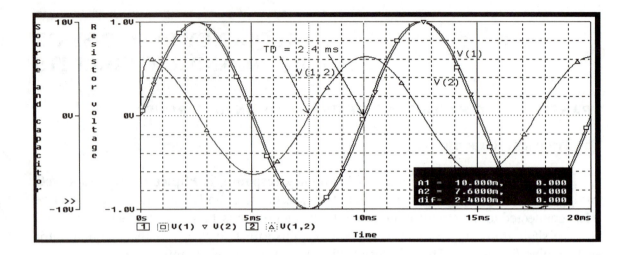

We shall next increase the frequency of the voltage source **V1** to 5 kHz. We need to change the **transient analysis** to 800 microseconds to allow the circuit to settle into its steady state. The results of the analysis are shown.

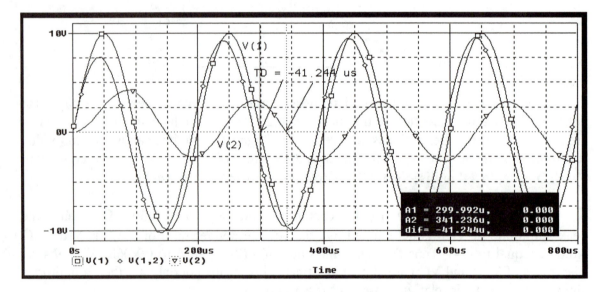

The capacitor voltage peak now measures 3.0 V. Also, the capacitor voltage now lags **V1** by a **TD** of − 41.244 us. This corresponds to a phase lag of 74.2°. The reactance of the capacitor at 5 kΩ has declined from 15.9 kΩ to 318 Ω. The decline of the capacitor voltage V(2) has resulted in an increase in the resistor voltage V(1,2). We can intuit that if we increase the frequency even further, ultimately the reactance of the capacitor will approach zero ohms, the capacitor voltage V(2) will be zero volts and the resistor voltage V(1,2) will be equal to the source voltage V(1).

To sum up our analysis, it has been demonstrated that the voltages and the current in the circuit are frequency dependent because of the inclusion of the capacitor in the circuit. We shall next investigate the behavior of this circuit over a wide range of frequencies.

The Frequency Analysis (AC Analysis)

What's new?

1. The use of AC voltage sources **VAC** and **VSRC**
2. The Phase **(P)** traces of ac voltages and currents

Although it is possible to obtain the amplitudes and the phase shifts of the resistor and capacitor voltages for any applied frequency, such a process is cumbersome and not necessary. We can get a global view of circuit behavior for a wide range of frequencies by performing a frequency analysis of this circuit. In the parlance of the **PSpice** program, such an analysis is referred to as an **AC analysis**. We shall be analyzing the circuit in Figure 9.02. The same resistor and capacitor as in Figure 9.01 will be used. We must replace the transient voltage source, **VSIN,** with an AC source **VAC.** Proceed as follows:

1. Click on the **Place part** icon.
2. In the **Place Part** box, click on **SOURCE** in **Libraries:** box.
3. Scroll to **VAC** instead of **VSIN**; click on **OK.**
4. Place the part in the usual manner.
5. Click on the default parameter **1Vac** to change voltage to desired value. You need not type ac. We set the source to 10 V.
6. The default value for **VAC** is **0Vdc**. Accept it if such is desired.

The completed circuit is shown next:

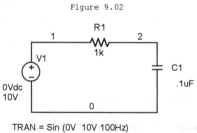

Figure 9.02

TRAN = Sin (0V 10V 100Hz)

We next set up the **AC analysis** as follows:

1. Click on **PSpice.**
2. Click on **New Simulation.**
3. Enter **1** as the name, click on **Create.**
4. In **Simulation Settings –1**, in analysis type, scroll to **AC Sweep/Noise**, click on it.

5. In **AC Sweep type**, click on **Logarithmic** to select it.
6. Scroll to **Decade**; click on it to select it.
7. In **Start Frequency**, click in the box to select it, type 1 Hz.
8. In **End Frequency**, click in the box to select it, type 100 kHz.
9. In **Points/Decade**, click in the box to select it, type 10.
10. Click on **OK** to accept all entrees.
11. Run the analysis.

The completed **Simulation Settings –1** dialog box, showing all entrees, is given next. The selection of a logarithmic rather than a linear scale allows for a wide range of data to be shown on the **PROBE** screen. The selection of the number of 10 points/decade has been found adequate by the author in most cases. The reader is, of course, at liberty to change the various parameters as desired.

Running the **AC analysis** of the circuit in Figure 9.02 produced the traces of the source voltage V(1), the resistor voltage V(1,2) and the capacitor voltage V(2).

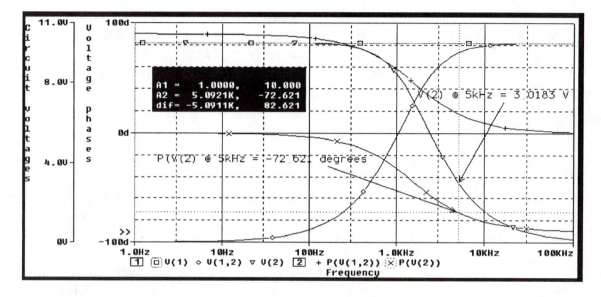

The source voltage **V1** has a constant peak amplitude of 10 volts and is independent of frequency. This is a desirable state of affairs. Such voltage could be obtained from a power supply. Much effort is devoted to insure that such a device delivers a constant voltage over a wide range of frequencies.

The capacitor voltage V(2) declines from 10 volts to zero volts, and the resistor voltage V(1,2) increases from zero volts to 10 volts as the frequency is changed from 1 Hz to 100 kHz. During that change of frequency, the phase angle of the capacitor voltage V(2) declines from 0° to -90° while the phase angle of the resistor voltage V(1,2) declines from 90° to 0°.

The Relationship Between the Transient Analysis and the AC Analysis

All these findings are in line with our previously stated intuitions based upon the **transient analyses** of this circuit carried out at 100 Hz and 5 kHz. We can see from the **PROBE** plots obtained by the **AC analysis** how the circuit behaves over a wide range of frequencies. By contrast, the **transient analysis** provided data at only one frequency at each run of **PSpice.** Let us now compare the data obtained from the two analysis types at the frequency of 5 kHz. Looking at the **PROBE** traces obtained from the **AC analysis**, we observe that at 5 kHz, the amplitude of the capacitor voltage V(2) is indicated as 3.0183 volts. This corresponds directly with the data from the **transient analysis** for that voltage. The phase angle of that voltage is indicated as –72.63 °. Again, this corresponds closely with the phase angle of –74.2° obtained from the **transient analysis**. The **AC analysis** gives the phase angles in degrees, so we do not need to convert from any **TD.**

In circuit analysis, traditionally we analyze circuit behavior often in either the **time domain (transient analysis** in **PSpice** jargon) or in the **frequency domain (AC analysis** in **PSpice** jargon). Often times, the two analysis types are separated by time. It is fortunate that the **PSpice** program allows us to perform the two analysis types separated only by a few keystrokes.

The VSRC Voltage Source

In the circuit of Figure 9.01 we used the **VSIN** voltage source to perform a **transient analysis.** This voltage source will not allow an **AC analysis.** In Figure 9.02 we used the **VAC** voltage source. This voltage source will not allow a **transient analysis.** **PSpice** has a voltage source labeled **VSRC** that allows the running of both a **transient** and an **AC analysis.** We turn next to its use.

We shall use the circuit in Figure 9.03a. It has the same resistor and capacitor as Figures 9.01 and 9.02. To select and place the voltage source **VSRC**, in the **Libraries** box of the **Place Part** dialog box, scroll to **SOURCE** and click on it. Either enter the part name or scroll to **VSRC**; it's located right after voltage source **VSIN**. Click on it, and click on **OK** to place it. The circuit will appear as shown.

Figure 9.03a

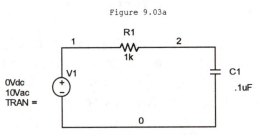

The first two parameters are the **AC analysis** parameters. We leave the offset voltage (0Vdc) at zero volts dc as is. The voltage amplitude (10Vac) of **VSRC** is changed from 1 volt to 10 volts. This will give us an **AC analysis** for the circuit in Figure 9.03a. **TRAN** pertains to the setting of the **Transient analysis** parameters. To set these parameters, proceed as follows:

1. Double click on the symbol of V1; the **OrCad Capture-[Property Editor]** screen opens.
2. On the bottom of that screen, click on **Parts** tab. It will be highlighted.
3. On **Schematic1:Page 1:V1,** scroll right to the **TRAN** column.
4. Enter **SIN(0V 10V 100Hz).** Don't forget the parentheses.

5. Click on **Display…;** the **Display Properties** dialog box opens.
6. Click on **Name and Value**, click on **OK**, and click on **Apply.**
7. Close the **Property Editor.**
8. Click on **V1** to deselect.

At the completion of this process, the circuit of Figure 9.3b will appear as shown:

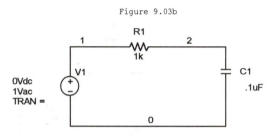

Figure 9.03b

The **AC parameters** are in order: a 0 V offset voltage and a peak voltage of 10 V. The **Transient parameters** in order are a sine wave with 0 V offset voltage, a 10 V peak voltage and a frequency of 100 Hz.

To run both the **transient** and the **AC analysis,** we proceed as follows:

1. Click on **PSpice**, click on **New Simulation Profile**. The **New Simulation** box opens.
2. In the **Name** field, type **1,** click on **Create,** the **Simulation Setting** dialog box opens.
3. In the **Analysis type** field, accept **Time Domain (Transient).**
4. Enter all **Transient parameters** as for Figure 9.01.
5. Click in the **Analysis type** field; a drop-down menu will appear.
6. Click on **AC Sweep/Noise** to select it.
7. In **Simulation Settings**, enter all **AC parameters** as for Figure 9.02.
8. Click on **OK.**
9. Click on **OK** and click on **Run**. The **AC analysis** will be run.
10. Obtain **PROBE** traces of the desired voltages. They should be identical to those in Figure 9.02.
11. At completion of the run, exit **PROBE** to return to the **Schematic page**.
12. Click on **PSpice**; click on **Edit Simulation Settings.**
13. Click in the **Analysis type** field.
14. From the drop-down menu, click on **Time Domain (Transient)** to select it.
15. Click on **OK.**
16. Click on **PSpice** and click on **Run**. The **Transient analysis** will be run.
17. Obtain **PROBE** traces of the desired voltages, they should be identical to those for Figures 9.01 and 9.02.

The Input Impedance of an RC Circuit

We noticed from our previous investigations of the **RC** circuits of Figures 9.01, 9.02 and 9.03a and 9.03b that in each case circuit behavior depended upon the applied frequency and the circuit elements, in our case a resistor and a capacitor. We shall next study in detail how these two circuit components are effected by frequency and how they determine the input impedance of a circuit. Let us turn to the circuit in Figure 9.04.

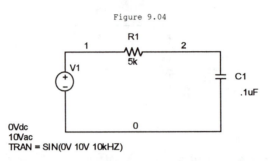

Figure 9.04

By definition, the input impedance is equal to the ratio of the source voltage divided by the source current. Given our nodal assignments, this ratio can be expressed as:

$$Zin = \frac{V(1)}{I(R1)} \Omega$$

In the formulation of complex algebra, this is equal to:

$$Zin = (R1 - jXc)\Omega$$

The first term is defined as the real part, and the second term, as the reactive part of the impedance. Let us attain the traces of each of them as a function of frequency. An **AC analysis ranging** from 1 Hz to 100 kHz produced the traces shown below. Incidentally, do not use 0 Hz for the lower limit. **PSpice** does not like zero.

The trace of the resistor **R1** is independent of the frequency range. It is a constant at 5 kΩ. This may sometimes not be true in real life. The reactance X_C of the capacitor tends toward infinity ohms at very low frequencies. This approximates an open circuit. As the frequency increases, X_C declines toward zero ohms. This approximates a short circuit. The input impedance **Zin** approximates the dominant X_C at low frequencies and approximates the dominant resistance **R1** at high frequencies. In sum, at low frequencies, **Zin** behaves like a capacitor. At high frequencies, **Zin** behaves like a resistor in the case of Figure 9.04.

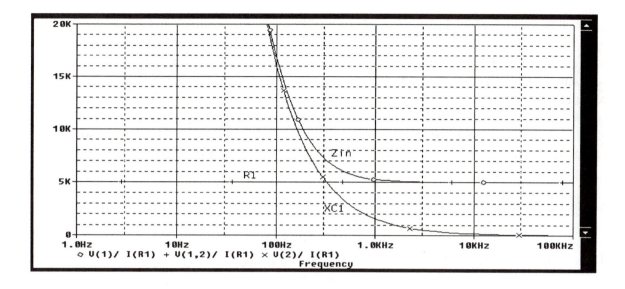

Let us next investigate the phase angle of **Zin** as a function of the applied frequency. To do this, we simply need to obtain the phase angle of the circuit current relative to the input voltage **V1.** This is done next.

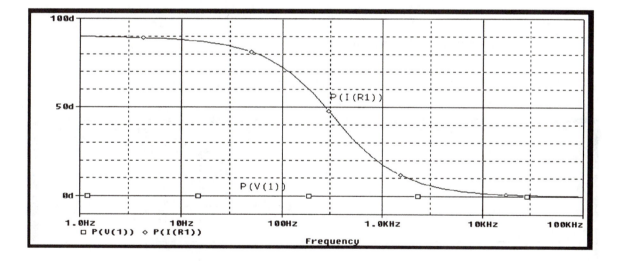

The trace of the phase of the circuit current I(R1) shows that at low frequencies, the current I(R1) leads the input voltage **V1,** or its nodal equivalent V(1) close to 90 °. This is due to the dominance of X_C at those frequencies. As the frequency increases, the phase angle of the circuit current I(R1) approaches 0°. This is due to the dominance of **R1** at those frequencies. The frequency discrimination of the **RC** circuit is made use of in circuit defined as filters. We turn next to that topic.

The Single-Section RC Circuit as a High-Pass Filter

A high-pass filter is a device that passes only signals that contain frequencies above a certain **critical frequency**, and it diminishes and ultimately rejects signals with frequencies below that **critical frequency**. The critical frequency is a design feature of the filter and is determined by the values of the resistor(s) and the capacitor(s) in the filter circuit. The region above the critical frequency is defined as the **passband region**, the area below the critical frequency is defined as the **reject region** of the filter.

The circuit in Figure 9.05 is in the high-pass configuration. The output voltage of the filter V(2), is taken across the resistor R1. We shall vary the frequency of the input voltage **V1** and note the resultant changes in the output voltage, the circuit current and the power in the circuit. Also, we shall investigate the **transient response** of this circuit at 10 kHz and relate our findings to the data from the **AC analysis**.

Figure 9.05

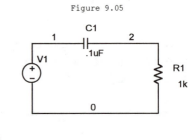

OVdc
10Vac
TRAN = SIN(0V 10V 10kHZ)

The circuit voltages, resulting from an **AC analysis**, are shown next.

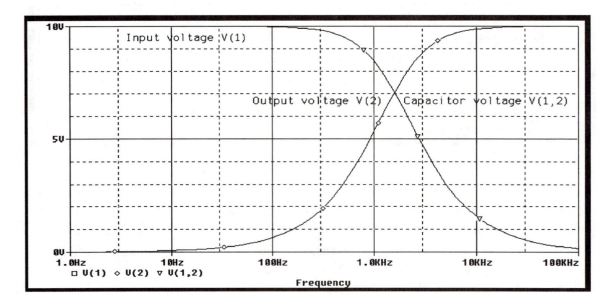

The source voltage **V1** remains constant at 10 volts. It is independent of frequency. Resistor voltage V(2) increases from zero volts toward the voltage of **V1**, while concurrently the capacitor voltage declines from its voltage equal to that of **V1** toward zero volts. We recall that the voltage across the resistor **R1** was defined as the output voltage, or the output signal, as it is often referred to in an electronic circuit. Now the question arises: above what voltage of V(2) are we in the **Passband region**, and at what voltage of V(2) are we in the **Reject region** of this filter? The convention has been adapted that the **critical frequency**, which is the dividing frequency between these two regions, is defined as that frequency at which the power to the resistor is one-half of the maximum power delivered to it by the source. We shall next determine its value both analytically and graphically.

The maximum power delivered to the resistor **R1** happens when the frequency is 100 kHz.

At that frequency:

$$P_{max} = \frac{(10)^2}{1000} = 100mW$$

Hence, at the **critical frequency**, the power to the resistor **R1** is equal to 50 milliwatts. From this

$$P_{(critical)} = 50mW = \frac{(V_2)^2}{1000}W$$

Solving for the resistor voltage V_2,

$$V_2\sqrt{(50mW)(1000)} = 7.07V$$

Thus, at the **critical frequency**, the voltage V(2) has increased to 7.07 volts. To determine the **critical frequency**, we need only until move a cursor along the trace of V(2) until it reads 7.07 volts. This was done, with the **critical frequency** shown to be 1.593 kHz.

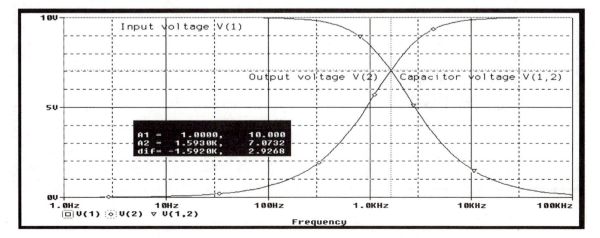

Observe that the **critical frequency** is located at the intersection of the capacitor with the resistor voltage. Since the **critical frequency** is, by definition, that frequency at which half the power of the voltage source is absorbed by the resistor R1, the other half is associated with the capacitor. This requires that XC1 = **R1**. This provides us with another way of determining the **critical frequency**. Study **the PROBE** traces shown next.

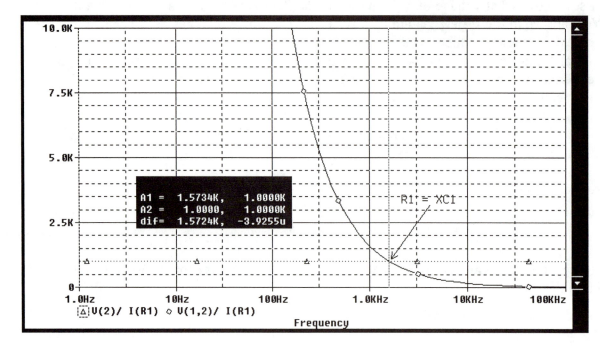

The traces of **R1** and of **XC1** intersect at the critical frequency.

We noted that at the **critical frequency**, the power to the resistor was equal to power received by the capacitor. Let us plot the power to both these elements.

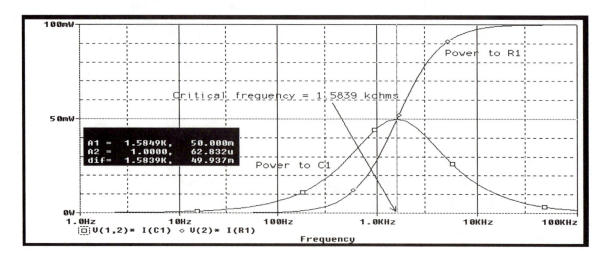

The intersection of the two power curves means that the power to **R1** is equal to the power to **C1**. This happens at the **critical frequency**. To obtain its value, we simply move one of the two cursors to the intersection of the power curves for the resistor and the capacitor, as shown above, and read the value of the critical frequency either from the **cursor coordinate box** or the X-axis, as shown.

We are not done yet with our efforts to determine the **critical frequency**.

Expressing Zin in complex form as was done above:

$$Zin = (R1 - jXc)\Omega$$

At the critical frequency **R1 = X$_C$**. Hence the impedance angle is:

$$\angle\theta = \arctan(X_C/R1) = 45°$$

This is confirmed by the **PROBE** trace, which shows the phase plot for the high-pass filter.

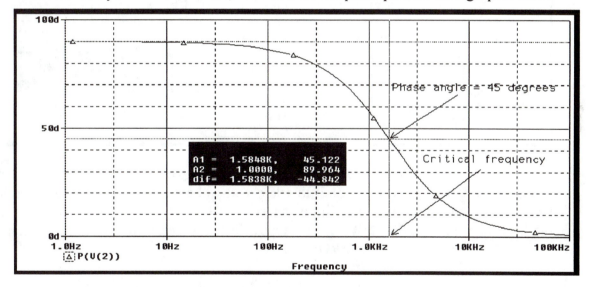

Voltage Gain and Phase Response of the High-Pass Filter

It is customary to express the voltage gain of a filter as the ratio of the output voltage divided by the input voltage. Many filter systems consist of several **RC** sections wired together, each having their own individual voltage gain. The overall voltage gain for such a system would be the product of the gains for each stage. However, if all the voltage gains were expressed as logarithmic ratios of output voltage divided by the input voltage, the overall voltage gain for the system would be the sum of the gains of each stage.

The decibel voltage gain ratio is defined as:

$$A_{VdB} = 20\log_{10}\left(\frac{V_2}{V_1}\right)dB$$

We can plot the decibel gain of a filter in **PROBE.** Let us obtain the numeric and the logarithmic voltage gains on two **Y-axes** next.

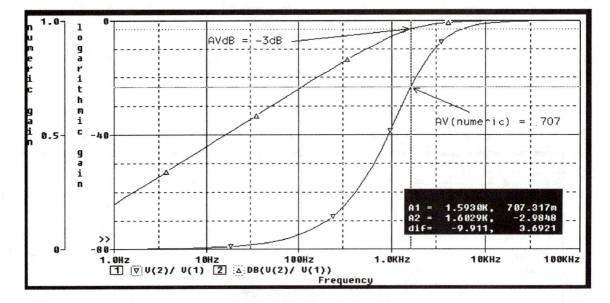

To obtain the **dB** gain, in the **Add Traces dialog box,** scroll to **DB( )** in the **Function or Macros** box. Click on it to select it. It will appear in the **Trace Expression** box. Get the ratio of V(2)/V(1). It will be added to **DB( ).** The completed expression for the logarithmic gain **DB(V(2)/V(1))** will appear in the **Trace Expression** box. Click on **OK** and the trace will appear on **PROBE** screen.

For the numerical voltage gain, we see the familiar S-curve shape. The range of that gain is from zero to unity. That latter gain means that the output voltage, V(2), is almost identical to the input voltage V(1). The decibel gain rises in an almost linear fashion from about – 70 dB to a maximum of zero dB. The logarithmic gain is never positive because the output voltage V(2) never exceeds the input voltage, as it would if we were investigating the output voltage of an amplifier system.

We observed before that at the **critical frequency**, the output voltage, V(2), was 7.07 volts. This translates into a numerical voltage gain of .707. Let us place one cursor on the trace of the numerical voltage gain of .707. The corresponding frequency is the **critical frequency**, 1.59 kHz. If we now use the second cursor on the logarithmic trace and line it up at the **critical frequency**, we obtain a corresponding logarithmic voltage gain of –3 dB. At the **critical frequency**, the **dB gain** is down –3 dB from its maximum value of 0 dB in our case.

The linear portion of the trace within the **Reject region** displays the roll-off rate of the filter circuit. It is a measure of the rate at which any frequency below the **critical frequency** is attenuated. For the single-stage circuit of Figure 9.05, the roll-off rate is **–20 dB/decade**. To verify this, the reader is encouraged to place one cursor at 1 kHz and the other cursor at 100 Hz on the trace of **DB(V(2)/V(1)).** The difference in **dB** will be **–20 dB,** as indicated on the logarithmic gain axis.

A Two Section RC High-Pass Filter

We extend our investigation into filter circuits by adding an identical **RC** section to the circuit in Figure 9.5 as is shown in Figure 9.6. The two **RC** sections need not be identical. Our objective is to find if there is any effect upon the roll-off rate of this two-section filter compared to the single section filter.

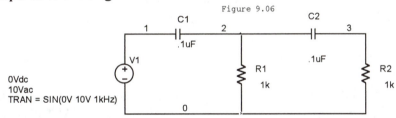

Figure 9.06

The result of the **AC analysis** is shown next.

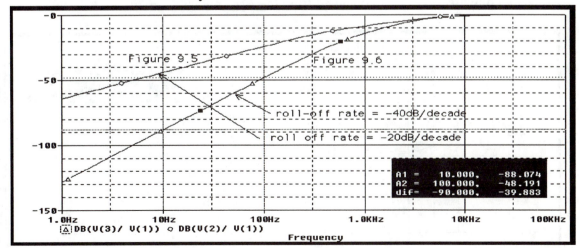

We observe that the roll-off rate for the overall filter comprising the two identical **RC** sections is steeper than that rate for the single-section filter. If we place our cursors on the trace of **DB(V(3)/V(1)),** we observe that for the decade from 100 Hz to 10 Hz, the logarithmic gain of the overall filter has changed by – 40 dB/decade. This is double the rate for the single-stage filter. In practical terms this means that the double-section filter discriminates more effectively against unwanted frequencies compared to the single-stage filter circuit.

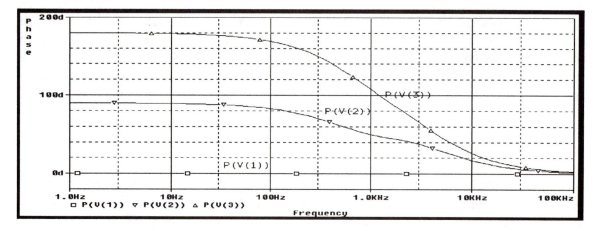

The phase plot for this filter is shown above. The results show that the addition of the second **RC** section has resulted in an overall increase of the phase shift from 90° at 1 Hz to 180° at that frequency. As the frequency is increased, eventually the phase shift between V(1) and V(3) will approach 0°, as it was for the single section filter.

Since we did set our **VSRC** to do a **transient analysis** at 1 kHz, let us do it next and compare its data with that from the **AC analysis** at 1 kHz. At 1 kHz, we are operating in the **reject region** of the filter. The results of the **transient analysis** are shown next. They are compared to the data obtained from the **AC analysis**.

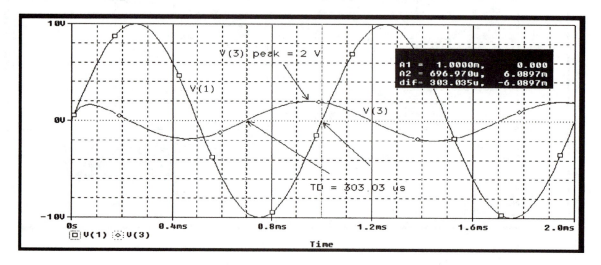

At the frequency of voltage source **V1** set to 1 kHz, the output voltage V(3) has a peak amplitude of 2 volts. It leads the input voltage V(1) by a **TD** of 303.03 microseconds. This corresponds to a phase lead of 109.2°

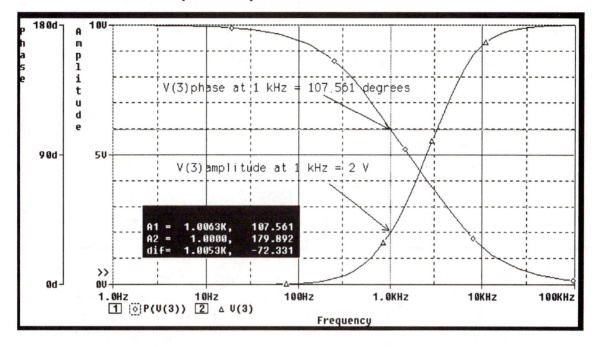

The **AC analysis** data is consistent with that from the **transient analysis**. At the frequency of 1 kHz, the amplitude of the voltage V(3) is 2 V.   That voltage has a phase shift of 107.561° that corresponds closely with that obtained from the **transient data**.

The RC Circuit as a Low-Pass Filter

By interchanging the position of the resistor and the capacitor in the previous circuits, we obtain a low-pass filter. The output voltage is now taken across the capacitor. We shall use the circuit in Figure 9.07 to study the properties of the low pass filter. We again use the **VSRC** voltage source to allow us to run both an **AC** and a **transient analysis**.

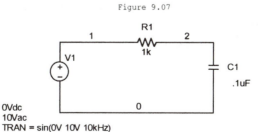

Figure 9.07

The **PROBE** traces of the input voltage **V1** and the output voltage V(2) are shown.

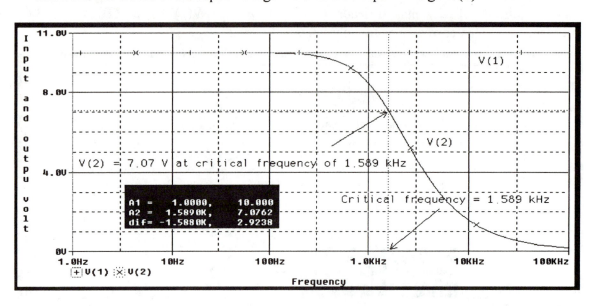

The trace of **V1** shows its independence from the changes in frequency while V(2) is a function of it. Specifically, the **passband** of this low-pass filter extends from 0 Hz up to the **critical frequency** of the filter. At that frequency, the output voltage has declined to .707 of its maximum voltage of 10 volts, or 7.1 volts. The **critical frequency** is 1.59 kHz. This is the same **critical frequency** as that of the high-pass filter. The reason for this is that in both cases, the values of **R1** and **C1** are identical. The value of the **critical frequency** can be obtained analytically by:

$$f_{(critical)} = \frac{1}{2\pi RC} \text{ Hz}$$

The **reject region** extends for any frequency larger than the **critical frequency** of this filter.

We next plot the traces of the **numerical** and the **logarithmic gain** of this filter. The numerical gain declines to .7 of its maximum value at the **critical frequency** of 1.59 kHz and the logarithmic gain declines by −3dB at that same frequency. The slight differences from their theoretical values are accounted for by the inability to place the gains at their precise theoretical values. The small differences are acceptable.

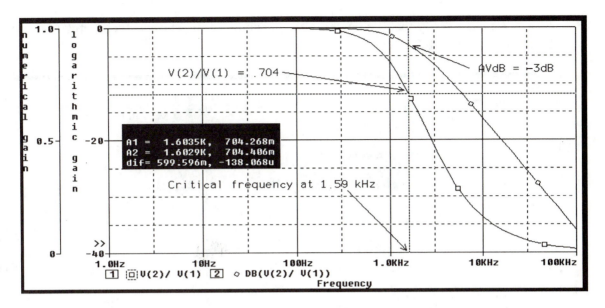

Next, we shall obtain the phase angle of the filter as a function of frequency.

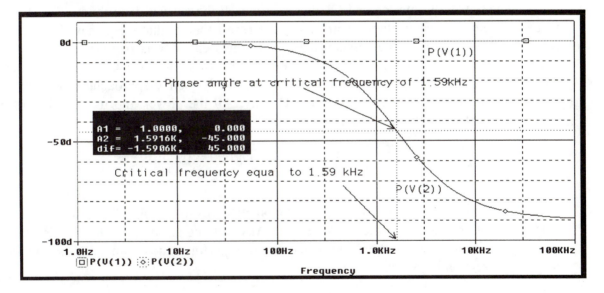

The trace of the phase angle shows that in the **pass region** of the filter, the phase angle changes from 0° to -45° at the **critical frequency** of 1.59 kHz. Increasing the frequency further, the filter operates in the **reject region** and the phase angle ultimately declines toward -90°. Thus the output voltage, V(2), starting out in phase with the source voltage, **V1,** at low frequencies, slides more and more behind the source voltage, **V1,** as the frequency is increased.

We shall verify that the power to the capacitor at the **critical frequency** is one-half of the total power to the filter.

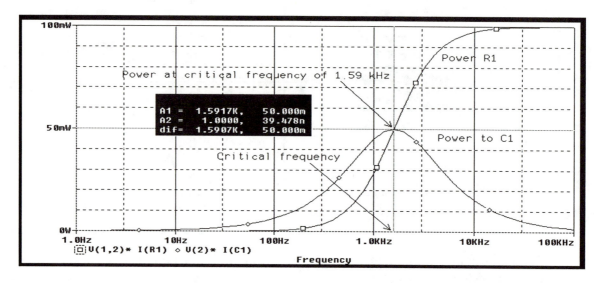

The power to both **R1** and **C1** increase until at the **critical frequency** of 1.59 kHz values are equal to each other at 50 mW. We recall that at that frequency the reactive reactance X_C is equal to the resistance **R1**. As the frequency is increased further, that reactance declines toward zero ohms. In effect, the capacitor becomes a short circuit increasingly unable to absorb any power. Thus, as the frequency approaches 100 kHz, the circuit in Figure 9.07 consists of resistor **R1** only. All the power into the circuit will be dissipated by **R1**. Its amplitude is easily verified:

$$P(\mathbf{R1}) = [V(1)]^2 /R1 = (10)^2/1000 = 100 \text{ mW}$$

The reader is encouraged to run a **Transient analysis** of this filter and compare its data with that obtained from the **AC analysis**. Also, adding an identical **RC** section to the filter circuit in Figure 9.07, the reader is requested to perform an **AC analysis** as was done for the two-section high-pass filter.

The Band-Pass Filter

The **pass region** of the high-pass filter was at and beyond its **critical frequency**. The **pass region** for the low-pass filter was below and up to its **critical frequency**. Combining these two filters, we obtain the band-pass filter. This filter will have a **lower critical frequency** determined by its high-pass section and an **upper critical frequency** determined by its low-pass section. The interval between these two **critical frequencies** is

defined as the **pass-band,** or **bandwidth,** of the filter. Figure 9.08 is the band pass filter
that we will analyze.

Figure 9.08

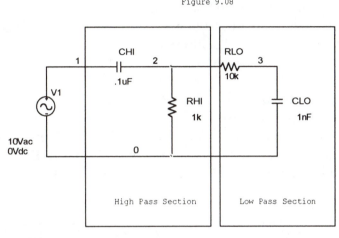

We perform an **AC analysis** ranging from 1 Hz to 100 kHz. The first **PROBE** plot
shows the traces of **V1** or its nodal equivalent V(1), the output voltage V(3) and its phase
angle relative to **V1**.

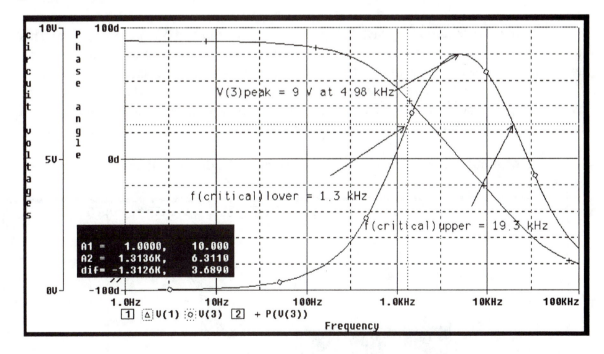

The output voltage V(3) reaches its peak voltage of 9 V at a frequency of 4.98 kHz.
Below that frequency, its amplitude is determined by the high-pass section of the filter,

while above that frequency, its amplitude is determined by the low-pass section of the band pass filter. The **lower** and **upper critical frequencies** are obtained by reducing the maximum amplitude of V(3) to .707 of its maximum voltage of 9 volts, or 6.3 volts. At that amplitude, the **lower critical frequency** is equal to 1.31 kHz and the **upper critical frequency** is equal to 19.3 kHz. The difference between the upper minus the lower critical frequency is defined as the **bandwidth** of the filter. This computes to:

$$Bandwidth = 19.3 \text{ kHz} - 1.3 \text{ kHz} = 18 \text{ kHz}$$

The frequency at which the voltage V(3) is a maximum is defined as the **center frequency** of the filter. It is indicated as 4.98 kHz on the above **PROBE** plot. The following mathematical relationship exists between the two **critical frequencies** and the **center frequency:**

$$f(center) = \sqrt{f(lowercutoff)(f(uppercutoff)} = \sqrt{(1.3kHz)(19.3kHz)} = 5kHz$$

The phase response below the **center frequency** is that of a high pass filter. The phase response above the **center frequency** is that of a low pass filter. The reader is referred to the phase responses of the high and low filters covered above to verify these two statements.

At the center frequency of 5 kHz, the voltages V(1) and V(3) are in phase. This makes the filter at that frequency equal to a resistor. At the **center frequency**, the reactance of the series capacitor **CHI** is 312 Ω. This reactance is small enough not to effect the operation of the circuit at the **center frequency**. At the **center frequency**, the reactance of the shunt capacitor **CLO** is 31.2 kΩ. This reactance is large enough not to effect the operation of the circuit at the **center frequency.**

To continue our analysis, we obtain the traces of the **numerical** and **logarithmic gains** of this filter.

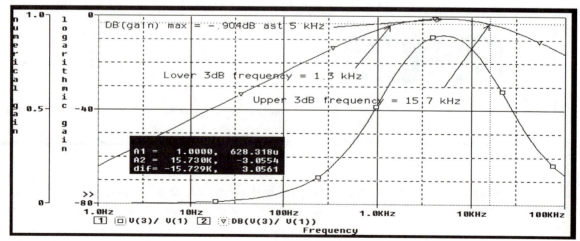

The **numerical** and the **decibel gains** have their maximum values at the center frequency of 5 kHz. To obtain the **critical frequencies**, move one cursor above and the other cursor below the **center frequency** to a point −3 dB below the maximum **dB** gain. Alternatively, on the **numerical gain** trace, move one cursor above and the other cursor below the center frequency to a point equal to .707 times the maximum numerical gain. From our **PROBE** plot above, we see that the **critical frequencies** are the same as those obtained from the voltage traces of this filter.

Frequency Responses of RC Circuit and RL Circuit Compared

We turn next to investigate the input impedance of an **RL** circuit. The fundamental difference between it and the **RC** circuit is that the reactance X_L, unlike the reactance X_C, is directly proportional to the applied frequency. The reactance X_L is defined as follows:

$$X_L = 2\pi fL \ \Omega$$

The input impedance Z_{in} of the **RL** circuit, in the formulation of complex algebra, is:

$$Z_{in} = (R1 + jX_L)\Omega$$

Again, the first term is the real part; the second term is the reactive part of the impedance. We shall run an **AC analysis** for the frequency interval from 1 Hz to 5 kHz of the circuit shown in Figure 9.09.

Figure 9.09

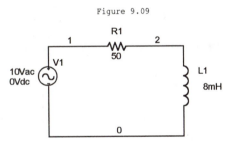

The resultant traces show that again, the resistor **R1** is independent of frequency. At low frequencies, since X_L is directly proportional to frequency, its reactance is very small. This effectively makes the inductor appear like a short circuit. Therefore, the impedance of Z_{in} is very nearly that of the resistor **R1**. As the frequency increases, so does X_L. It becomes ever more dominant as Z_{in} approaches it in value. At 5 kHz, the two are almost identical.

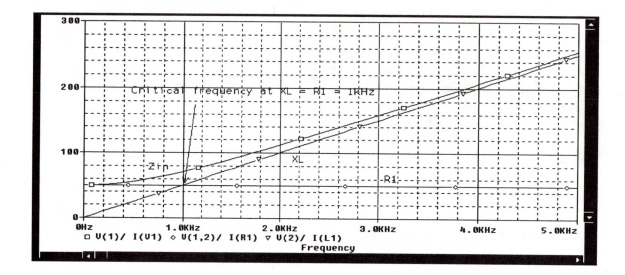

Since **XL** is a linear function of frequency, The X-axis was put into linear, not logarithmic, mode. In the **Axis Settings**: dialog box, click on **Linear** in the **Scale** box. Although the **AC analysis** was run for 100 kHz, the **X-axis** was changed to range of from zero hertz to 5 kHz. This was done to show more of the low frequency portion of our traces. The **critical frequency**, as it was in the case of the **RC** circuit, is found at the intercept of the traces of **R1** and **XL**. It is determined from a cursor placed at the intercept as 1 kHz.

We next obtain the circuit voltages of this circuit.

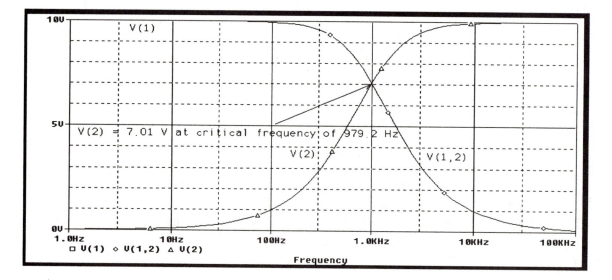

As was the case for the **RC** circuit, the equivalency of the inductive reactance with the resistance of **R1** makes the traces of the inductor and the resistor voltage intercept at the **critical frequency**. We observe that at that frequency, the amplitude of both voltages is .707 times the voltage of the source, **V1,** or 7.1 volts. This corresponds to the case of the **RC** circuit.

Our next **PROBE** plot shows the phases of the input voltage V(1) and that of the output voltage V(2).

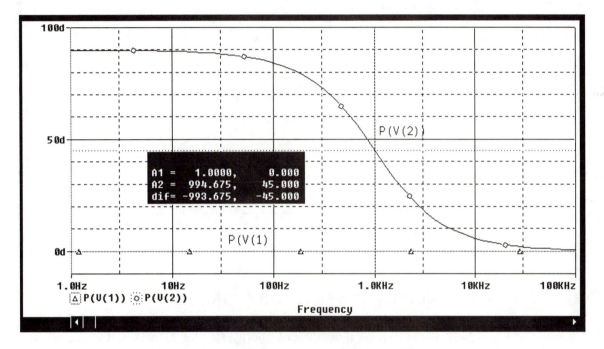

While V(1) is in phase with itself, at low frequencies, the phase angle of V(2) is close to 90°. As the frequency increases to the **critical frequency**, the phase angle now is 45°. Any further increase of the frequency results in the phase angle of V(2) declining toward zero degrees. If we were to observe these two waveforms on an oscilloscope, we would see that at low frequencies, these two voltages are a quarter of a wavelength apart. However, with an increase of frequency, the waves would slide into each other.

We shall next investigate the power flow in this circuit. We shall find that there is a similarity in that power flow compared to that of the **RC** circuit and there is an important dissimilarity. The traces of the power to **R1** and **L1** are shown next.

For the **RC** circuit, at low frequencies, the capacitor is an open circuit and consequently no power can flow in the circuit. For the **RL** circuit, at low frequencies, the inductor is a short circuit and thus maximum power flows to the resistor **R1**. As the frequency increases, the capacitive reactance declines while the inductive reactance increases. Ultimately, as we get to 100 kHz, the capacitor is a short circuit and all the power goes to **R1**. At that frequency, the inductor is an open circuit, and all power stops flowing in the circuit. If we turned the power traces of the two circuits end over end, they would be identical. No doubt the reader has noticed the analogous behavior of the **RC** and the **RL** circuits. It is for this reason that the application of the **RL** circuit as a filter is not covered. The reader by now has sufficient material to proceed on this topic on his/her own.

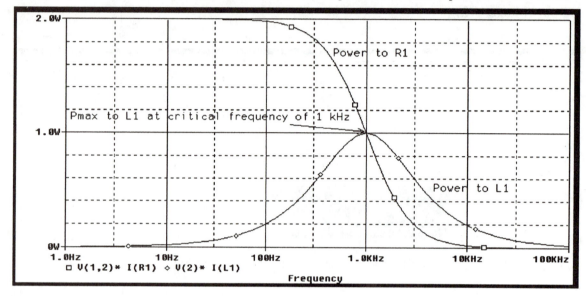

FREQUENCY (AC ANALYSIS) OF A SERIES RLC CIRCUIT

An **RLC** circuit contains two independent energy-storing elements: the capacitor stores electrical energy and the inductor stores magnetic energy. If we apply a sinusoidal voltage source to a circuit containing these two elements, we will observe an energy exchange between them. The rate of that exchange is determined by the **natural resonance frequency** of the circuit. The capacitance and the inductance of the two reactive elements determine its frequency.

If the frequency of the applied current or voltage source is equal to the **natural resonance frequency** of the circuit, we have the phenomena of resonance. At that frequency, the reactive power received by one of the reactive elements is exactly equal to the reactive power send to it by the other. At that frequency, these elements neither receive nor do they send power back to the energy source. Thus, from the perspective of the energy source, the circuit consists of a pure resistor receiving power from that source.

The input impedance of the series **RLC** circuit is

$$Z_{(input)} = R + j(X_L - X_C)\Omega$$

At resonance. $X_L = X_C$, from which follows that the resonance frequency is:

$$f_{(resonance)} = \frac{1}{2\pi\sqrt{LC}} Hz$$

Since at resonance $Z_{(input)} = R$, the circuit current will have its maximum amplitude determined as follows. It will also be in phase with an applied voltage source.

$$I = \frac{Vsource}{R} amps$$

We shall begin our investigation of the **RLC** circuit with the circuit shown in Figure 9.10. An **AC analysis** will be performed from 100 Hz to 20 kHz. To obtain smooth traces, the points per decade setting was increased to 100 points/decade.

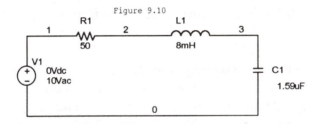

Figure 9.10

The traces of the circuit voltages are shown first.

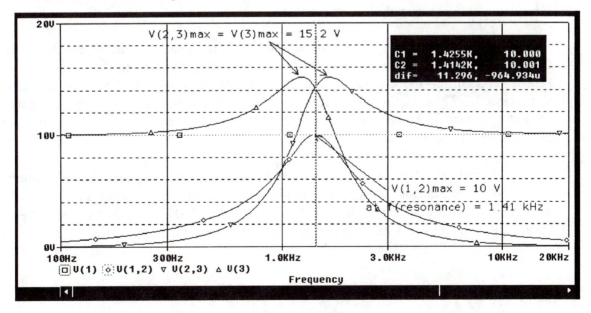

To render the interpretations of the voltage traces more intelligently, Figure 9.11 shows the operating conditions of the circuit in Figure 9.10 at the indicated frequencies.

Figure 9.11: Circuit in Figure 9.10 at various frequencies

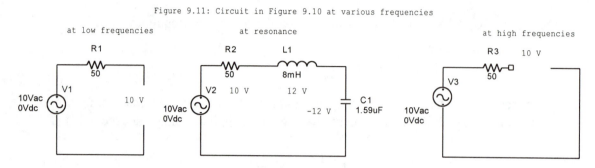

At low frequencies, the capacitor is effectively an open circuit and the inductor effectively a short circuit. No current can flow in the circuit. Thus, the resistor voltage V(1,2) and the inductor voltage V(2,3) are equal to zero volts, and the entire source voltage **V1** is across that capacitor.

As the frequency of the source is increased, the reactance of the capacitor declines and that of the inductor rises. A current begins to flow, a voltage develop across all three-circuit elements.

At the **resonance frequency**, the resistor voltage V(1,2) is at its maximum of 10 volts. This voltage is equal to the source voltage. The inductor voltage V(2,3) and the capacitor voltage V(3) are both equal in their absolute amplitudes However, these two voltages are 180° out of phase with each other. Their sum consequently is equal to zero volts. Hence, from the perspective of the source, the circuit in Figure 9.10 is reduced to **R1**.

A further increase in the frequency results in a further decline in the reactance of the capacitor and an increase in the reactance of the inductor. When the frequency of 20 kHz is reached, the inductor effectively has become an open circuit and the capacitor a short circuit. Thus, again, no current can flow. The resistor voltage V(1,2) and the capacitor voltage V(3) are equal to zero volts, and the source voltage appears across the inductor terminals, as shown.

The voltage traces allow for a visual determination of the **resonance frequency**. It occurs when the resistor voltage V(1,2) reaches its maximum and when the traces of the inductor and capacitor voltage intercept. From our **PROBE** plot this happens when the frequency is equal to 1.41 kHz. We shall verify its value analytically:

$$f_{(resonance)} = \frac{1}{2\pi\sqrt{LC}} = \frac{1}{2(3.14)\sqrt{(8mH)(1.59uF)}} = 1.41kHz$$

Oh, how nicely it all works out.

The traces shown above give only the absolute value of the amplitudes of the circuit voltages. To further gain insight into the behavior of this circuit, we shall obtain the traces of their phases next.

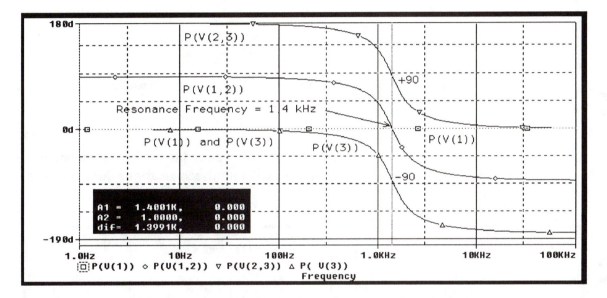

The phase of the source voltage V(1) remains at zero degrees over the range of the frequencies of the AC analysis. In plain English, that voltage has no choice but to be in phase with itself.

The phase of the resistor voltage V(1,2) starts out at low frequencies at a phase angle of 90° because of the dominance of the capacitive reactance at those frequencies. As the frequency is increased toward the **resonance frequency**, its phase angle declines because of the increasing equality of the capacitive and the inductive reactances. At resonance, they are equal to each other. In consequence, the circuit is a pure resistor with a phase angle of 0°. A further increase of the frequency makes the inductive reactance the dominant one and the phase angle of V(1,2) declines to -90°.

As we study the relative phases of the reactive voltages V(2,3) and V(3), we note that at any frequency, they are 180° apart. At the **resonance frequency**, the phase of the

the inductor voltage V(2,3) is 90°, whereas that of the capacitor voltage V(3) is at -90°. Thus, their relative phase is 180°.

Let us for a moment go into our electronic laboratory, use a **VSRC** voltage source and hardwire our circuit, connect it to a signal generator, use a multitrace oscilloscope, connect one channel across the signal generator and one channel each across the resistor, the capacitor and the inductor. Set the amplitude of the signal generator to 10 volts and its frequency to 1.4 kHz. Our hardwired circuit will be looking like this:

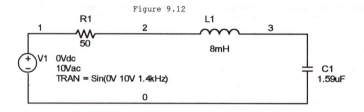

Figure 9.12

Since the oscilloscope is a **Time Domain** device, we need to perform a **transient analysis** on the circuit in Figure 9.12. The author used 2 milliseconds for its duration. The voltage traces are shown next.

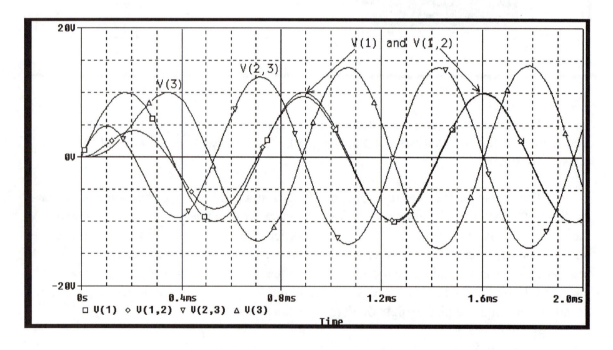

The source voltage V(1) and the resistor voltage V(1,2) are practically equal in amplitude and in phase at the resonance frequency. This agrees with our **AC analysis** data above.

The two reactive voltages are equal in amplitude but 180° out of relative phase. Also, their amplitude is larger that that of either the source or the resistor voltages. Again, this data is consistent with our **AC analysis** data.

All that data can, and has been seen from the **AC analysis** data above. This data is somewhat more abstract than that from the **transient analysis**. The latter data can be seen with the naked eye. The **AC analysis** data is not so obvious and needs interpretation in explaining what actually happens in a circuit. Our efforts must be directed to use both sets of data in an effective and complementary way.

Frequency Selectivity of the Series RLC Circuit

We observed before that at the resonance frequency of 1.41 kHz of the circuit Figure 9.12, the impedance of the circuit was a pure resistor and had a minimum value of 50 Ω. Consequently, the current in the circuit ought to be a maximum. Its trace shown next confirms this.

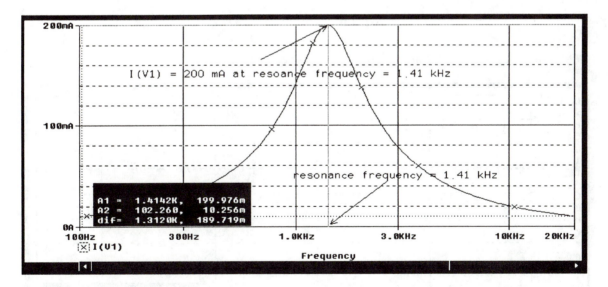

At the resonance frequency, the value of the current is equal to the voltage of V(1) divided by resistor R1. In symbols:

$$I(V1)_{resonance} = V(1)/R1 = 10 \text{ V}/50 \text{ } \Omega = 200 \text{ mA}.$$

As we depart from the **resonance frequency** in either direction, the current declines and the impedance of the circuit increases. Thus the circuit discriminates against certain frequencies. This means that this circuit can be used as a filter. We recall from our previous discussions of the low-pass and high pass-filters that the introduction of the half power concept resulted in defining a **pass region** and **rejection region** for these filters. We shall now extend the concept of half power to the **RLC** circuit.

At resonance, the circuit current is 200 mA and the circuit impedance is equal to 50 Ω. The power into the circuit at resonance is equal to:

$$P_{resonance} = I^2_{resonance} Z_{input} = (200mA)^2 * 50\Omega = 2watts$$

To determine the current at the half-power point(s), we solve the following equation:

$$I_{halfpower} = \sqrt{\frac{1}{50}} = 141.4mA$$

At the half-power point, the ratio of the current divided by the current at resonance is

$$141.4 \ mA/200 \ mA = .707$$

Let us next set our cursors at the half-power current.

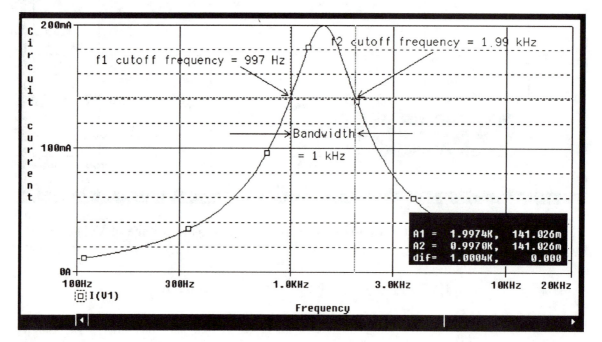

The current I(V1) reaches its predicted value of 200 mA at the resonance frequency of 1.421 kHz. As we move cursors **A1** and **A2** to a point of the trace when I(V1) is equal to 141.1 mA, we obtain an upper cutoff frequency f_2 equal to 1.99 kHz and a lower cutoff frequency f_1 equal to 997 Hz. The difference between these two frequencies, $f_2 - f_1$, equal to 1kHz, is defined as the **bandwidth** of this circuit. Our circuit in effect is a **bandpass filter**. Let us obtain its logarithmic gain and roll-off rate next.

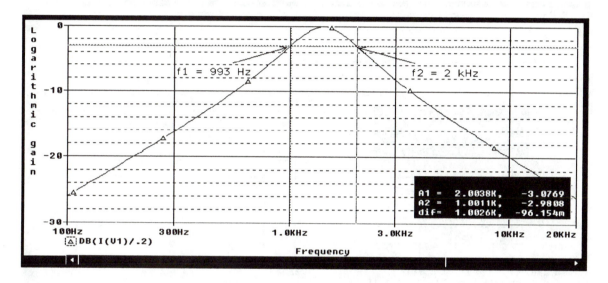

To obtain that trace, the current I(V1) was divided by its maximum value of 200 mA. At a gain of –3 dB, the critical frequencies were those obtained previously. The roll-off rate for this filter is **–23 dB per decade** as can be obtained from the plot of the trace as the frequency changes from 1 kHz to 100 Hz. This means that this circuit discriminates somewhat more effectively against frequencies out of the **pass band** than did the previously studied filter circuits.

The Quality Factor Q

Historically, the term **quality factor** derived from the ability of coils used in early radios to discriminate against unwanted frequencies when a radio was tuned to a particular station. After all, Beethoven and rap are hard to listen to at the same time. A high Q coil was referred to as one that let you listen to the one without hearing the other.

Mathematically, the **quality factor Q** for a series resonant circuit is defined as the ratio of the reactive power of the inductor or the capacitor in a circuit divided by the **RMS** power to the resistor in a circuit. At the **resonance frequency**,

$$Q = \frac{I^2 Xc}{I^2 R} = \frac{I^2 Xl}{I^2 R} = \frac{Xc}{R} = \frac{Xl}{R} = \frac{V(3)}{V(1,2)} = \frac{V(2,3)}{V(1,2)}$$

For our circuit, this ratio reduces itself as shown to either V(2,3)/V(1,2) or V(3)/V(1,2), depending upon whether we used the power to inductor or the capacitor in that ratio. Let us plot the traces of both of these ratios.

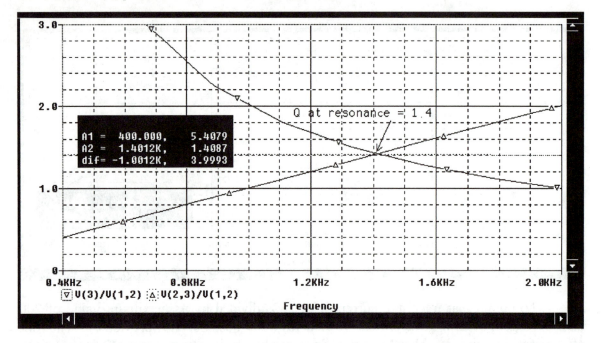

The traces show that at the **resonance frequency** of 1.4 kHz, and only at that frequency, both traces intercept. The **quality factor** is equal to 1.4. We noted before that at the resonance frequency, the power to both reactive elements was equal in amplitude. The intercept of the above traces is one more proof of that. The trace based on the inductor varies proportionately with frequency, whereas that based on the capacitor varies in inverse proportion.

As a proof of the correctness of our finding, we shall obtain the value of **Q** directly from the circuit components.

$$Q(resonance) = \frac{1}{R}\sqrt{\frac{L}{C}} = \frac{1}{50}\sqrt{\frac{8mH}{1.59uF}} = 1.4$$

Both determinations of Q agree.

Q and the Reactive Voltages at Resonance

At resonance, the total impedance into the circuit is the 50 Ω of resistor **R1**. To obtain the voltage across either the capacitor or the inductor, we use voltage division:

$$V_C = \frac{X_C}{R}V(1) \qquad\qquad V_L = \frac{X_L}{R}V(1)$$

But, at resonance:

$$\frac{X_C}{R} = \frac{X_L}{R} = Q = 1.4$$

Therefore:

$$V_C = 1.4V(1) \text{ volts} \quad \text{and} \quad V_L = 1.4V(1) \text{ volts}$$

This shows that at resonance, both reactive voltages are 1.4 times the source voltage **V1** or its nodal equivalent V(1). This makes their amplitudes equal to 14 volts. Thus, the circuit exhibits a voltage gain across both reactive elements.

Relationship Between Bandwidth and Q

The following relationship exists between the bandwidth and the quality factor Q at the resonance condition of the circuit:

$$bandwidth = \frac{f_{(resonance)}}{Q}$$

For the circuit in Figure 9.11, solving for Q at resonance:

$$Q = \frac{f_{(resonance)}}{(f_2 - f_1)} = \frac{1.4kHz}{(1.9kHz - .9kHz)} = 1.4$$

The Effect of Changing the Resistance R1 in Figure 9.11

From the various expressions given for **Q**, we note that it is inversely proportional to the resistance of our **RLC** circuit in Figure 9.12. We shall next change the value of R1 from 10 Ω to 90 Ω in steps of 20 Ω. Aside from studying the effect this has on circuit performance, it will give us the opportunity to review the **Parametric analysis**. For a review of this topic, the reader is referred to Chapter 1, pages 27-30. Figure 9.12 is reproduced as Figure 9.13.

Figure 9.13

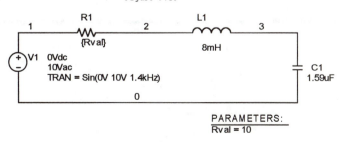

After the parameters of **R1** are set:

1.   Click on **PSpice** and click on **Edit Simulation Settings**

2.   Accept the **AC Sweep/ Noise** parameters shown below set by the previous run.

| General | Analysis | Include Files | Libraries | Stimulus | Options | Data Collection | Probe Window |

Analysis type:

AC Sweep/Noise ▼

Options:

☑ General Settings
☐ Monte Carlo/Worst Case
☑ Parametric Sweep
☐ Temperature (Sweep)
☐ Save Bias Point
☐ Load Bias Point

─ AC Sweep Type ─

○ Linear

◉ Logarithmic

Decade ▼

Start Frequency: 100

End Frequency: 100kHz

Points/Decade: 100

─ Noise Analysis ─

☐ Enabled

Output Voltage:

I/V Source:

Interval:

OK   Cancel   Apply   Help

3.   In the **Options:** box, click on **Parametric Sweep** to select it.

4.   Enter the parameters as shown next.

| General | Analysis | Include Files | Libraries | Stimulus | Options | Data Collection | Probe Window |

Analysis type:

AC Sweep/Noise ▼

Options:

☑ General Settings
☐ Monte Carlo/Worst Case
■ Parametric Sweep
☐ Temperature (Sweep)
☐ Save Bias Point
☐ Load Bias Point

─ Sweep variable ─

○ Voltage source   Name:

○ Current source   Model type: ▼

◉ Global parameter

○ Model parameter   Model name:

○ Temperature   Parameter name: Rval

─ Sweep type ─

◉ Linear   Start value: 10

○ Logarithmic Decade ▼   End value: 90

                      Increment: 20

○ Value list

5. Click on **OK**, click on **PSpice** and click on **Run.**
6. In the **Available Sections** box, click on **All** and click on **OK.**
7. Click on **Trace**. In the **Add Traces box and** in **Functions or Macros** box, select desired variables and functions.
8. Click on **OK** to plot the traces.

The Circuit Current I(V1)

The resultant traces of the numerical values of the circuit current I(V1) are shown for each of the values of **R1**.

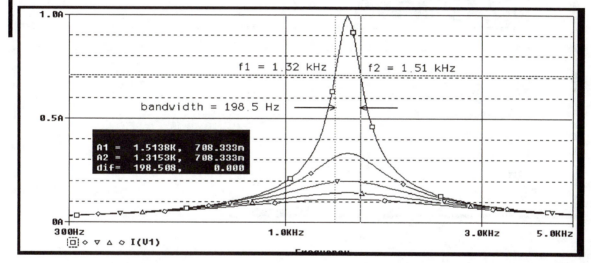

The maximum amplitude of I(V1) of 1 A occurs when **R1** is at its minimum value of 10 Ω. The minimum value is 200 mA happens when **R1** is equal to 50 Ω. We notice a sharp rise of the amplitude of I(V1) with decreasing **R1** and a decrease of the bandwidth of this circuit. This means that the circuit provides for greater amplification of its voltages and sharper discrimination against adjacent frequencies either side of its bandwidth with decreasing **R1**.

The Resistor Voltage V(1,2)

The traces of the resistor voltage V(1,2), the inductor voltage V(2,3) and the capacitor voltage are shown on separate plots below to avoid visual clutter. Let us first study the resistor voltage V(1,2). It may at first seem surprising that regardless of the different resistance values of **R1**, the resistor voltage V(1,2) remains at 10 volts. Let us recall that at resonance, the algebraic sum of the capacitor plus the inductor voltage is equal to zero volts. As the resistor **R1** changes its resistance, the **resonance frequency** does not change. This forces the voltage V(1,2) to remain at 10 volts. It is the circuit current that must adjust. At **R1** equal to 10 Ω, the current I(R1) must be 1 ampere so that V(1,2) remains 10 volts. At **R1** equal to 90 Ω, the current I(R1) changes to .11 ampere. This current maintains 10 volts across **R1**. This analysis holds true for any intermediate resistance values of **R1.** We also note that the lower the resistance of **R1**, the narrower is the bandwidth of this circuit. The cursors **A1** and **A2** were placed on the 10 Ω trace of V(1,2). It shows a bandwidth of 198.5 Hz. This is the same value as found for the traces of I(V1) above.

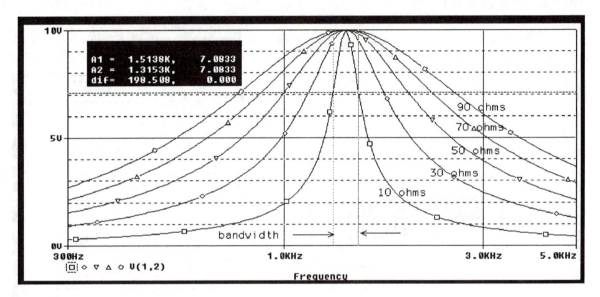

The Capacitor and Inductor Voltages

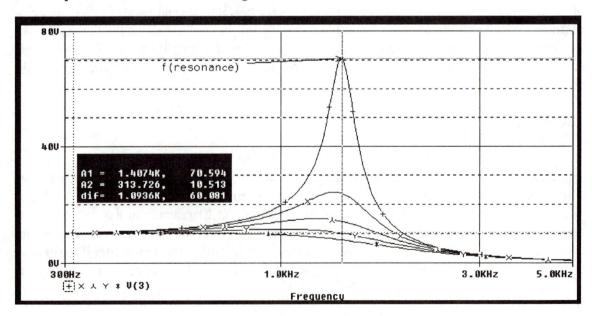

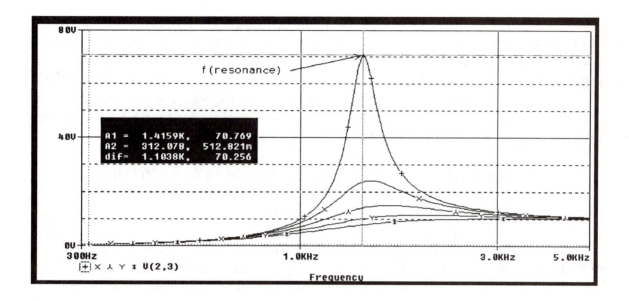

The traces of both these voltages are shown above on separate **PROBE** screens. Both voltage traces reach a peak voltage of about 70 volts at the **resonance frequency**. This voltage compares to about 14 volts when the resistance of **R1** was 50 Ω. Thus, with decreasing **R1**, higher voltage peaks of the reactive voltage were obtained. Indeed, we recall that the reactive voltages are equal in amplitude to the product of the voltage of the source **V1** times the **Q** of the circuit. From this follows that the resonant **Q** with **R1** being equal to 10 Ω is equal to 70/10 = 7.

Q (resonance) as a Function of R1

Lastly we shall investigate the changes in the resonant **Q** as **R1** changes in accordance with the parametric analysis.

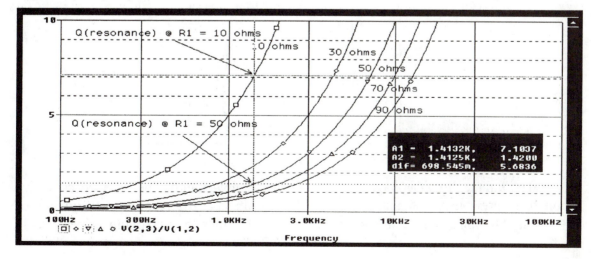

The traces show the variations in **Q** as the resistance of **R1** changes over the range of indicated frequencies. Cursor **A1** was placed at the intercept of the resonance frequency with the **Q** trace when **R1** was equal to 10 Ω. The indicated value of Q was 7.1. This was predicted. Cursor **A2** was placed at the intercept of the **resonance frequency** with the **Q** trace when **R1** was equal to 50 Ω. The indicated value of **Q** was 1.4. This is consistent with its earlier obtained value. We calculated a fivefold increase for **Q** as **R1** was changed from 50 Ω to 10 Ω. The latest PROBE plot confirms this.

A note on procedure: To obtain the position of cursor **A1** on the 10 Ω trace, depress the cursor icon. Five symbols appear below the **X-axis**, each corresponding to a run of the **parametric analysis**. Click the right mouse key on the first one, it will be highlighted. Use the right arrow key to place cursor **A1** into desired position along the trace. Next, right click on the third symbol from the left. It corresponds to the third run of the **parametric analysis** with **R1** equal to 50 Ω. Simultaneously depress the **SHIFT** key and the right arrow key. Move cursor **A2** into desired position along that trace. Looking at the row of symbols for **Q** on the **PROBE** plot, the reader will notice that the left most and the third symbol from the left are enclosed within a small square.

FREQUENCY (AC ANALYSIS) OF A PARALLEL RLC CIRCUIT

All the concepts introduced in the analysis of the series RLC circuit also pertain to the analysis of a parallel **RLC** circuit, with some modifications. Thus, we shall somewhat shorten our discussion of the analysis of the parallel **RLC** circuit in comparison to that of the series **RLC** circuit. We shall analyze the circuit in Figure 9.14. We perform an **AC analysis** from 100 kHz to 900 kHz and obtained the data shown next.

Figure 9.14

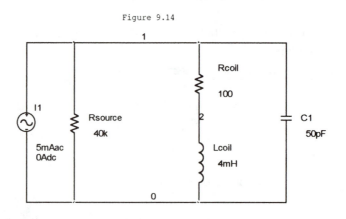

The unique circuit voltage V(1) shows the by now familiar bell-shaped curve we encountered previously. Again, its peak of 190.1 volts is at **the resonance frequency** of 354.4 kHz. Dropping our cursors to positions of .707 times the maximum value of V(1)

allows us to determine the critical frequencies f_2 equal to 400.6 kHz and f_1 equal to 317 kHz. This yields a bandwidth of 83.6 kHz.

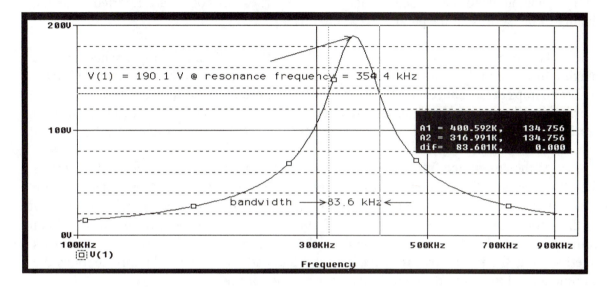

The next **PROBE** plot shows the amplitudes of the source current and the three branch currents in the circuit.

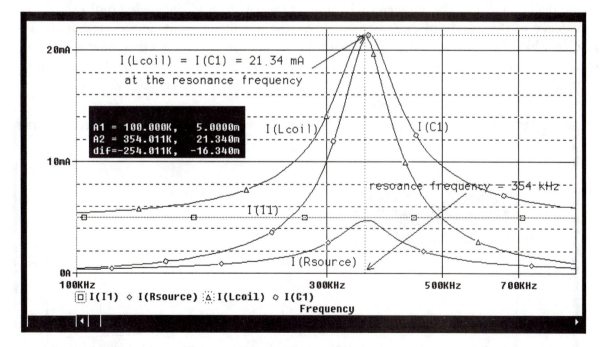

The shapes of these traces look similar in shape to those encountered previously. In particular, if we compare them with those of the voltages traces obtained for Figure 9.10 on page 297, we observe that the source current I(I1) is analogous to the source voltage

V(1). The resistor current I(Rsource) is analogous to the resistor voltage V(1,2). The inductor current I(Lcoil) is analogous to the capacitor voltage V(3). The capacitor current I(C1) is analogous to the inductor voltage V(2,3). These analogies are referred to as the duality of the series and the parallel **RLC** circuits.

The reactive currents reach an amplitude of 21.34 milliamps at resonance, and the maximum value of the resistor current I(Rsource) is also equal to the source current I(I1) of 5 milliamps at resonance. This is of interest. The current source I(I1) delivers 5 mA into the circuit, all of which flows through **Rsource** at resonance. Yet, the two reactive currents are 21.34 milliamps. Their ratio to that of the source current computes to 4.27.

At the **resonance frequency**, the current **I(Rsource)** develops a 190 V drop across the 40 kΩ resistor **Rsource** and across **Lcoil** and **C1**. At resonance, their reactances are both 8.9 kΩ. The 100 Ω resistor of **Rcoil** is swamped by the inductive reactance of 8.9 kΩ of the Lcoil. Dividing the voltage V(1) of 190 V by either of them results in a current equal to 21.3 mA. This current circulates through **Lcoil** and **C1** only. This current is referred to as the **tank current** of the circuit. The current source I(I1) only supplies the current needed to make up the powerloss in resistor **Rsource**.

If we compare the phase of the currents in the **parallel RLC** circuit shown below with the voltage phase of the **series RLC** circuit shown on page 299 we see further evidence of the duality of these two circuits. The reactive currents in the present circuit behave like the reactive voltages in the former circuit. In both cases they have a relative phase shift of 180 Ω between them. We recall that the algebraic sum of the reactive voltages was equal to zero volts. In the present circuit, the algebraic sum of the two reactive currents also is zero amperes.

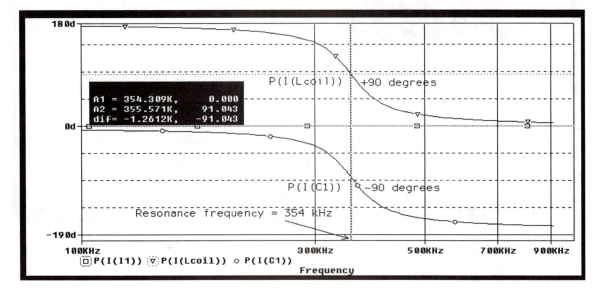

The quality factor **Q** for this circuit is defined as the ratio of the reactive current divided by the source current. Thus, from our data for this circuit we have

$$Q_{(resonance)} = \frac{I(C1)}{I(I1)} = \frac{21.3mA}{5mA} = 4.26$$

The value of **Q** as a function of frequency is shown on the next trace. In the analytic expression for **Q** or in that for the trace, the inductor current I(Lcoil) could have been used.

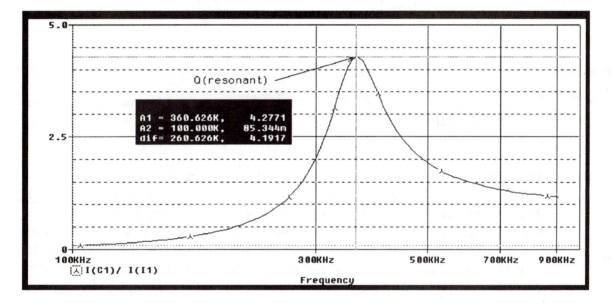

PROBLEMS

9.1   For this circuit, perform a **transient analysis** with the frequency of the voltage source **V1** set to 100 Hz. Also, perform an **AC analysis** ranging from 1 Hz to 10 kHz. From the data obtained from them, answer the following
       (a). What is the voltage of V(2) at 100 Hz from the **AC analysis** data?
       (b). What is the voltage of V(2) from the **transient analysis** at that frequency?
       (c). Compare the two amplitudes.
       (d). Obtain the **TD** and the corresponding phase shift between V(1) and V(2) from
            your **transient analysis** data at 100 Hz.
       (e). At that frequency, what is that phase shift from your **AC analysis** data?
       (f). Compare the two phase shifts obtained.
       (g). Determine the **critical frequency** of this circuit.
       (h). What is the amplitude of V(2) at that frequency?

(i). With the frequency of the **transient analysis** set to 100 Hz, is the circuit operating in the **pass region** or the **reject region**?

(j). What is the **roll-off rate** of this filter?

(k). How would you classify this filter circuit?

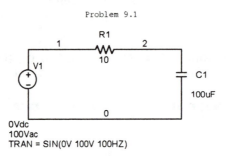

Problem 9.1

9.2   Using the same circuit as in Problem 9.1, in addition to the **AC analysis**, perform a **transient analysis** at 300 Hz. Answer all questions as in Problem 9.1.

9.3   The circuit shown in this problem is that of a compensated oscilloscope probe. Before we begin its analysis, can you predict its behavior at low and high frequencies? Recall that the reactance of a capacitor is inversely proportional to the applied frequency. For a compensated probe, the product **R1C1 = R2C2**. This makes the output voltage V(2), which is the voltage to the scope, insensitive to frequency. Run an **AC analysis** of this circuit from 1 Hz to 100 kHz to show that insensitivity. What is the percentage variation of V(2) over the range of frequencies investigated?

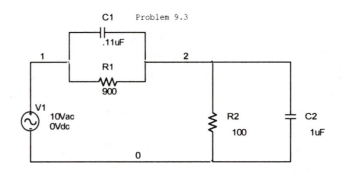

Problem 9.3

9.4   For the high-pass filter shown, determine the critical frequency by any of the methods of Chapter 9. After you have obtained it, increase it by a factor of 10. To do this, divide the value of the capacitor by 10. The resistor value remains the same. This procedure is defined as **frequency scaling**. Run an **AC analysis** with the new value of the capacitor. Verify the shift in the **critical frequency**.

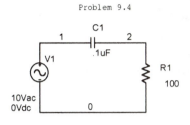

Problem 9.4

9.5   For the multi section **RC high-pass** filter, find the **critical frequency** of the filter. Obtain both the numerical and the logarithmic gains as a function of frequency. What is the **roll-off rate** for this filter?

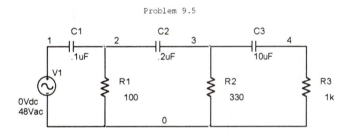

Problem 9.5

9.6   For the circuit shown, find the power to the resistor at the **critical frequency**. Determine the watt rating of that resistor so that the filter can safely operate in its **passband**.

Problem 9.6

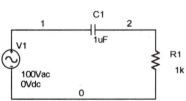

9.7   It is desired to design the high-pass filter shown that starts its **passband** at 16 Hz. At the **critical frequency**, the power to the resistor is 1 watt. Using these design criteria, determine the needed voltage of source **V1** and the capacitor.

Problem 9.7

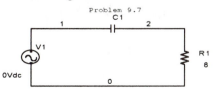

9.8   Perform an **AC analysis** for range of frequencies from 1 Hz to 100 kHz and find
         the following:
         (a). The amplitude of the input impedance at 1 kHz
         (b). The phase angle of V(4) at 1 kHz
         (c). All nodal voltages
         (d). The current through each circuit element
         (e). The **critical frequency**
         (f). The **roll-off rate** in dB/decade
         (g). The gain of dB(V(4)/V(1)) from 100 Hz to 1 kHz
         (h). The power to the circuit for the range of frequencies investigated.

Problem 9.8

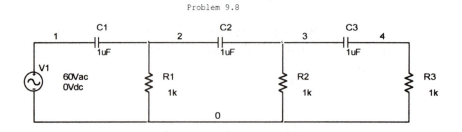

9.9   For this filter circuit, from an **AC analysis**
         (a). Obtain the intercept of the capacitor with the resistor voltage.
         (b). At what frequency do they intercept and why?
         (c). Obtain the phases of the capacitor with the resistor voltage.
         (d). What is their phase shift at 1 Hz?
         (e). What is their phase shift at 10 kHz?
         (f). What is their relative phase shift over the range of frequencies used in the **AC
              analysis?**
         (g). At what frequency are the two phase shifts of equal magnitude? To help you
              answer that question, form the expression P(V(1,2)) + P(V(2)).
         (h). What are the maximum numerical and logarithmic gains? V(2) is the output
              voltage and V(1) is the input voltage.
         (i). What are the minimum numerical and logarithmic gains?
         (j). What are these two gains at the **critical frequency**?
         (k). What is the **roll-off rate** for this circuit?
         (l). What is the maximum power input to the circuit and at what frequency?
         (m). What is the maximum power to the resistor and at what frequency?
         (n). What is the maximum power to the capacitor and at what frequency?
         (o). At what frequency is the power to the capacitor equal to the power to the
              resistor?
         (p). What is the value of that power?

(q). How does this power compare to that obtained for the resistor in question (m)?

Perform a **transient analysis** at 1 kHz and answer the following questions:
(a). What are the amplitudes of V(1), V(1,2) and V(2)?
(b). Compare these amplitudes to those obtained from the **AC analysis** at 1 kHz.
(c). What are the relative phase shifts between these voltages?
(d). Compare these phase shifts to those obtained from the **AC analysis** at 1 kHz.
(e). Find the **RMS** power to the resistor.
(f). Compare that power with the power to the resistor from the **AC analysis** at 1 kHz.

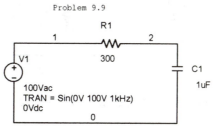

Problem 9.9

9.10   You are given the circuit diagram and the traces of the power to the resistor and the capacitor.

Problem 9.10

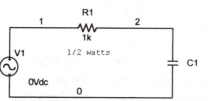

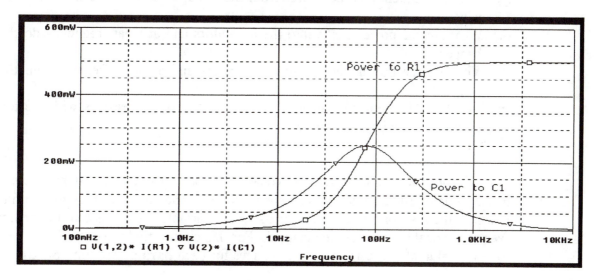

From it, design a circuit so that the maximum power to the resistor does not exceed .5 watts in the pass band. Determine the capacitance of the capacitor, the maximum allowable voltage of **V1** and the critical frequency of this circuit.

9.11   The circuit in this problem is a band-pass filter.

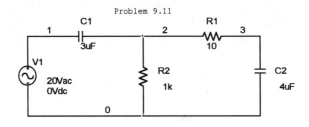

(a). With V(3) considered the output voltage, what is the numerical gain of this filter?

(b). Obtain the dB gain of this filter.

(c). What are the maximum numerical and logarithmic gains in the passband?

(d). From these gains, determine the corner frequencies f_1 and f_2.

(e). What is the **bandwidth** of this filter?

(f). Determine the **Q** of this filter.

(g). What is the phase of V(3) relative to **V1** at the corner frequencies? Use 2 Y-axes to answer this question.

(h). What are the **roll-off rates** of this filter?

(i). What is the maximum power into the filter?

(j). At what frequency is the power from the voltage source **V1** a maximum and why?

(k). What are the power peaks into the capacitors and at what frequency do they happen?

(l). What are the power peaks into the resistors and at what frequency do they happen?

(m). At 100 kHz, compare the sum of the powers to the resistors with the total power delivered by the source.

9.12   The **RL filter** shown has a variable resistor that can change its resistance from an initial value of 100 Ω to 900 Ω in steps of 200 Ω. The output voltage for the filter is across the resistor. The reason for using the variable resistor is that the **critical frequency** of this filter can be changed. For this circuit, find the critical frequency for each value of the resistance as it changes for each **parametric** run. Specify the maximum change for the **critical frequencies** as the resistor changes from its minimum to its maximum resistance.

Problem 9.12

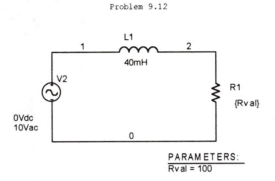

9.13   The circuit in Problem 9.12 had its **AC** voltage source replaced by **V2**, which is a **VSRC** voltage source. It is desired to run a **transient analysis** at 1 kHz. It is important to relate the data of both the **AC analysis** and the **transient analysis** for the voltage V(2) to insure that they are consistent with each other. Compare the two sets of data to insure that this is so.

Problem 9.13

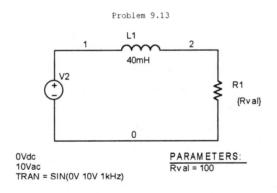

9.14   For this circuit, determine the **critical frequency** from a trace of the filter output voltage V(3). At that frequency, what is the power to **R2** and to **L2**? Compare the power from the source **V1** to the power delivered to **R1** and **R2** in the **pass region** of this filter. On separate Y-axes; obtain the numerical and the logarithmic gains of this filter. Compare the **roll-off rates** for the two gains. Are they equivalent?

Problem 9.14

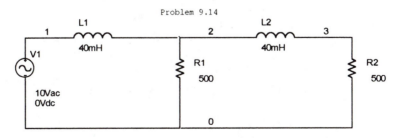

9.15   The circuit shown in this problem is a model of a tuned amplifier. Its job is to amplify a signal of a specified frequency while rejecting signals containing different frequencies. The circuit contains a voltage-dependent current source with a gain of 4 mS.

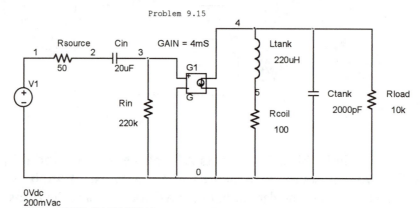

Problem 9.15

0Vdc
200mVac
TRAN = SIN(0V .2V 237.65KHz)

For this amplifier:

(a). Find the output voltage V(4) as a function of frequency. Set the X-axis for a range of frequencies from 10 kHz to 10 MHz.

(b). What is the maximum voltage of V(4)?

(c). What is the **resonance frequency** of the amplifier?

(d). From the trace of V(4) determine the corner frequencies f_2 and f_1.

(e). What is the **bandwidth** of the amplifier?

(f). Determine the **Q** of the circuit from your frequency measurements.

(g). Verify that $QV(1) = V(4)$.

(h). Obtain the numerical and the logarithmic gain of this filter.

(i). Verify that these two gains are consistent with each other.

(j). It is next desired to obtain the voltages V(1) and V(4) as functions of time. This means we perform a **transient analysis**. Its parameters are set as shown by the **VSRC** source. Perform this analysis for 20 microseconds and obtain the traces of V(1) and V(4). In particular, what is the phase shift between these two voltages?

9.16

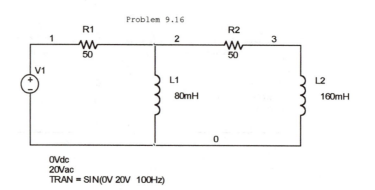

Problem 9.16

0Vdc
20Vac
TRAN = SIN(0V 20V 100Hz)

Perform an **AC analysis.**
(a). Get the traces of all the circuit voltages on one **PROBE** plot.
(b). What are the inductor voltages at the low frequencies?
(c). What are the inductor voltages at the high frequencies?
(d). What are the resistor voltages at the low frequencies?
(e). What are the resistor voltages at the high frequencies?
(f). Draw an equivalent circuit at the low and one at the high frequencies.
(g). From these equivalent circuits, explain your findings above.
(h). Classify this filter.
(i). Obtain the **critical frequency**.
(j). Obtain the **roll-off rate**.
(k). Obtain the numerical and the logarithmic gain of the ratio of V(3)/V(1).
(l). Are the gains consistent with each other?
(m). What is the phase angle of V(3) at 100 Hz?

Perform a **transient analysis.**
(a). Obtain the traces of V(1) and V(3) as functions of time.
(b). What is the **TD** and the corresponding phase angle between them?
(c). Compare the amplitude of V(3) obtained by the **transient analysis** with that obtained from the **AC analysis.** Are they the same?

9.17 For this filter, first perform an **AC analysis** from 100 Hz to 20 kHz.

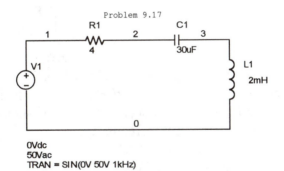

OVdc
50Vac
TRAN = SIN(0V 50V 1kHz)

(a). Obtain traces of the circuit voltages.
(b). What are these voltages at 100 Hz; at 20 kHz?
(c). Draw an equivalent circuit at 100 Hz and at 20 kHz.
(d). Use the equivalent circuits to explain your findings above.
(e). What is the **resonant frequency**?
(f). Get the **corner frequencies f_1 and f_2**.
(g). What is the **bandwidth** of this filter?
(h). What is the **roll-off rate**?
(i). What is its **Q**?
(j). Using two Y-axes, obtain the traces of V(3) and P(V(3)).

(k). What is the phase of V(3) at 1 kHz?

Next, perform a **Transient analysis** at 1 kHz.
(a). Obtain a trace of the voltages V(1) and V(3).
(b). Compare their amplitudes with those obtained from the **AC analysis.**
(c). What are their relative **TD** and phase angle?
(d). Compare this data with that from the **AC analysis.**
(e). At 1 kHz, is the circuit operating within or outside its **passband?**

9.18   The filter in this problem is a band-stop filter. Thus, the frequency interval between
       the two **cutoff frequencies** is defined as the **stopband**, and the frequencies outside
       these two frequencies define the two **passbands** of this filter. The voltage V(2) is
       defined as the output voltage.

Problem 9.18

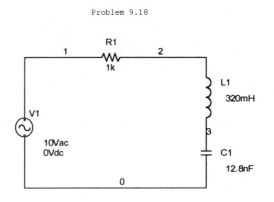

Perform an AC **analysis** ranging from 500 Hz to 10 kHz.
(a). Plot the traces of V(2) and its phase P(V(2)) on the same **PROBE** plot
(b). What is the minimum voltage of V(2) at the **resonance frequency?**
(c). What is the phase of V(2) at that frequency?
(d). Obtain the **corner frequencies**.
(e). Determine **the stopband**.
(f). Obtain the numerical and the logarithmic gains. Use two Y-axes.
(g). What are the values of their minimum gains?
(h). Are they consistent with each other?

9.19   The circuit in Problem 9.18 has been modified by the addition of the variable resistor **R2** connected across the output terminals of the filter. It is desired to study any possible changes in the performance of this filter as **R2** changes its resistance from 100 Ω to 1900 Ω in steps of 300 Ω. Perform an **AC analysis** for this modified filter from 1 kHz to 10 kHz. Summarize any changes in performance of this filter compared to that in Problem 9.18.

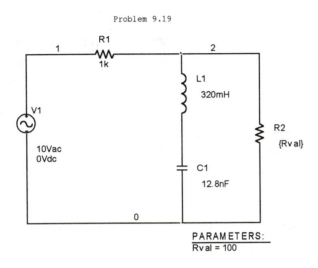

Problem 9.19

PARAMETERS:
Rval = 100

9.20   For this filter, perform an **AC analysis** to answer the following questions. The output voltage is across **Rload**. Note: The frequency range for that analysis is to be determined by the reader.

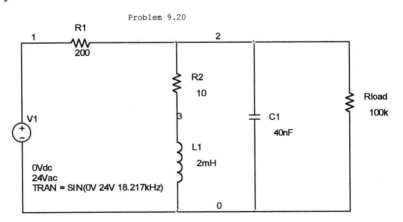

Problem 9.20

(a). Using two Y-axes, obtain the traces of V(2) and of P(V(2)).
(b). From this data, what is the maximum of the output voltage V(2) of this filter?
(c). What is the **resonant frequency**?
(d). At that frequency, what is the phase angle of V(2)?

(e). Determine the corner frequencies f_1 and f_2.
(f). What is the **bandwidth**?
(g). Determine the **Q** for this filter.
(h). Obtain the numerical and the logarithmic gain.
(i). Do they correspond with each other?
(j). Determine the **roll-off rate**.

9.21   For the circuit in Problem 9.20 perform a **transient analysis** at the indicated frequency of the **VSRC** source. Show two cycles of the traces of V(1) and V(2). The result of your analysis points to an important way in which to use an oscilloscope to determine the **resonance frequency** of such a filter. Vary the frequency of a signal generator attached to this circuit until the two voltages V(1) and V(2) have no relative phase angle as seen on the oscilloscope screen. Now look at the dial of the signal generator: the indicated frequency is the **resonant frequency** of the circuit.

9.22   A 50 Ω load resistor is supplied by the two voltage sources shown.

Problem 9.22

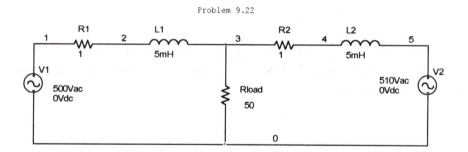

For this circuit find:
(a). The voltage across Rload from 100 Hz to 100 kHz
(b). The phase of that voltage for the same frequency interval
(c). The power to **Rload**
(d). The **critical frequency**
(e). The voltage V(3) at the **critical frequency**
(f). The power to **Rload** at the critical frequency
(g). The numerical and the logarithmic gains
(g). The **roll-off rate** of the filter
(h). What is the **dB** gain at 30 kHz?

9.23   For the circuit in Problem 9.23 perform an **AC analysis**, with the range of frequencies used in the analysis left to the reader. The voltage across **Rload** is the output voltage.

Problem 9.23

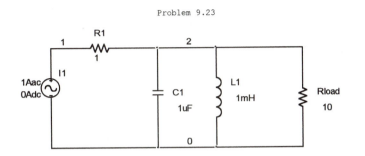

(a). Find the voltage V(2) as a function of frequency.
(b). Find the **resonance frequency.**
(c). Find the **corner frequencies.**
(d). Determine the **bandwidth**.
(e). Compute the **Q** for this circuit.
(f). Obtain the numerical and the logarithmic gains.
(g). What is the **roll-off rate** of this circuit?
(h). What kind of filter circuit is this?

9.24   Repeat Problem 9.23 for the circuit shown.

Problem 9.24

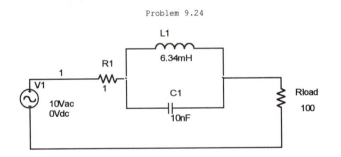

9.25   For the circuit shown, find the frequency at which the power to **Rload** is at a maximum. Determine the value of that power.

Problem 9.25

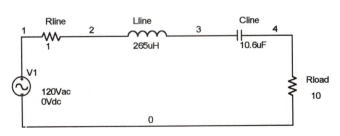

10

CIRCUITS WITH NON-SINUSOIDAL SOURCES

INTRODUCTION

In this chapter, we shall study the application of nonsinusoidal but periodic current and voltage waveforms to circuits containing capacitors, inductors and resistors. We shall show that these nonsinusoidal but periodic waves contain a multitude of frequency components. They are defined as the harmonic components of these waves. Their combination uniquely determines the shape of the current and voltage waveforms.

If such a wave is applied to a purely resistive circuit, all frequency components will be proportionally changed in amplitude. Thus, the wave shape, altered in scale only, will be preserved. If a circuit does have capacitors and/or inductors in addition to resistors, each frequency component will be acted on differently by the circuit, resulting in a distortion of the original waveform.

The **Fourier series** gives mathematical expression to the frequency content of a current or voltage waveform. We shall obtain its **RMS** value and the contribution that each harmonic component makes to the power to a load. Given a particular **Fourier spectrum**, we shall synthesize its corresponding time function. We shall also derive the **Fourier series** of some standard time functions and compare our results with those published in the professional literature. Finally, we shall obtain the response of both resistive and complex circuits to nonsinusoidal currents and voltages using **Fourier analysis**.

FREQUENCY COMPONENT OF A SQUARE PULSE

We stated that a periodic but nonsinusoidal current or voltage could be represented by a number of currents or voltages each at a different frequency. In theory, the number of such component frequencies is infinite. In practice, we can adequately synthesize a particular waveform with about 10 to 15 harmonics.

Let us begin by introducing the various sine waves that make up the square wave shown. The circuit in Figure 10.01 was used to obtain that square pulse.

Figure 10.01

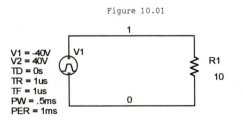

V1 = -40V
V2 = 40V
TD = 0s
TR = 1us
TF = 1us
PW = .5ms
PER = 1ms

V1

1

R1

10

0

The voltage pulse V(1) is shown next.

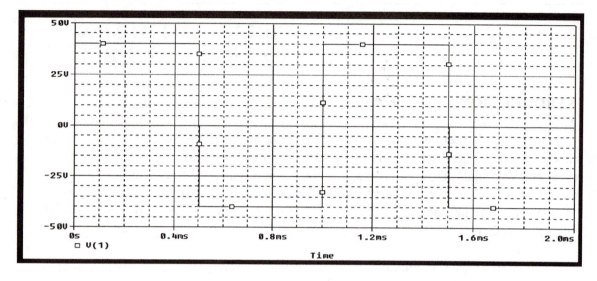

The circuit in Figure 10.02 has been constructed to show how the above square pulse can be synthesized by a series of sinusoidal voltages.

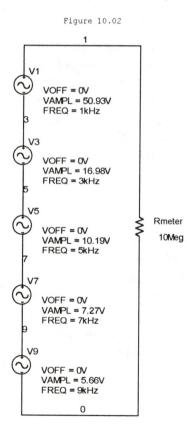

It shows five voltage generators that generate the first five harmonics of a series needed to synthesize the square pulse. At this point, we need not to concern ourselves with their individual amplitudes. It will be made clear why each of them had a particular one.

The result of running a **Transient analysis** for 2 milliseconds is shown below. The plot contains the voltages of each harmonic voltage generator in addition to the voltage V(1) across Rmeter. That resistor is the model of a digital voltmeter. The voltage V(1) across it is the sum of the five harmonic voltages. There are some important points to be made.

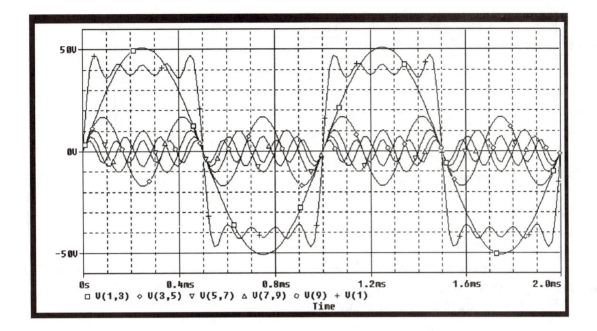

The voltage generated by **V1** had a frequency of 1 kHz. It is defined as the **fundamental frequency** of the square pulse. Observe that it is the largest of the harmonics. In fact, it has a larger amplitude than the square pulse. It determines the pulse width of the square pulse. In this regard, observe that all harmonics go through zero volts at .5 millisecond intervals, two of which determine the period of the pulse.

The next largest harmonic is the third one. As the order of the harmonic increases, the harmonics diminish in amplitude. Their frequencies are integral multiples of the fundamental frequency and their periods are equal to 1/**n** times the period of the fundamental voltage sine wave. In this ratio, **n** is the order of the **n**th harmonic.

V(1) is the sum of the first five harmonics. Already it can be seen that V(1) resembles a square pulse. We can intuit that if more harmonics were added to this summation, the resultant voltage V(1) would more nearly resemble a perfect square pulse In the following table is a summary of the harmonic content of our analysis. In it, **f1** represents the **fundamental frequency** and **T1**, the fundamental period.

Harmonic Component	Frequency	Period
Fundamental	f1	T1
3rd Harmonic	3f1	1/3T1
5th Harmonic	5f1	1/5T1
7th Harmonic	7f1	1/7T1
9th Harmonic	9f1	1/9T1

Fourier Theorem and Series

What's new?

1. The **Fourier Transform** in **PSpice**

The person who conveyed the rules of how to combine harmonic sine waves properly to synthesize a wave form such as the square pulse was Charles Fourier (1772-1837). This rule is stated as the Fourier theorem:

A periodic function $v(t)$ may be represented as an infinite series of terms such that

$$v(t) = A_o + A_1 cos\omega_0 t + A_2 cos2\omega_0 t + A_3 cos3\omega_0 t + \ldots + A_n cosn\omega_0 t$$
$$+ B_1 sin\omega_0 t + B_2 sin2\omega_0 t + B_3 sin3\omega_0 t + \ldots + B_n sin_n\omega_0 t$$

In this statement of the theorem, A_0 is the average, or dc, value of a periodic time function. It may or may not be zero for a particular function. Putting the symbols into words: an average value plus an infinite number of sine and cosine terms may represent a periodic function. The frequencies of these terms are harmonically related, that is, whole number multiples, of the fundamental radian frequency ω_0. In practice, 10 to 15 terms are usually adequate to synthesize most common waveforms.

The formulation above is known as the **trigonometric representation** of the **Fourier series**. It is neither the only one possible nor the most convenient in practice. An alternative form of the theorem is

$$v(t) = C_0 + C_n sin(\omega_0 t + \theta_n) \; volts$$

This is the form used by the **PSpice** program. The first term is the average value of the series.   The relationship between the coefficient C for each harmonic, and the corresponding cosine and sine coefficients is as follows:

$$C_n = \sqrt{A_n^2 + B_n^2} \qquad \qquad \theta_n = tan^{-1}\left(\frac{A_n}{B_n}\right)$$

An Application of the Fourier Series Theorem

Our objective is to obtain the harmonic content of the square pulse of Figure 10.01. The **PSpice** program will resolve it by means of a **Fourier series analysis** in conjunction with a **transient analysis**. The duration of the **transient analysis** must be at least equal to one period. In the case of Figure 10.01, this equates to 1 millisecond. To perform the **Fourier analysis**, proceed as follows: After the completion of the circuit diagram

1.   Click on **PSpice.**
2.   Click on the **New Simulation Profile** box.
3.   In **Name** box, enter 1 or any other convenient number; click on **OK.**
4.   In the **Simulations Settings** box, in **Analysis type**, select **Time Domain(Transient).**
5.   In **Run to time**, enter 2ms.
6.   In **Maximum step size,** enter .02ms.
7.   Click on **Output File Options . . .** . The **Transient Output File Options** dialog box opens.
8.   Enter the **Fourier parameters** as shown. Note that the center frequency is the fundamental frequency of our square pulse. Click on **OK.**
9.   In the **Simulation Settings** box, click on **OK.**

```
┌────────────────────────────────────────────────────────┐
│ Transient Output File Options                      [X] │
│                                                        │
│                                              OK         │
│   Print values in the output file every: │2ms  │ seconds │
│   ☑ Perform Fourier Analysis              Cancel        │
│      Center Frequency:    │1kHz │  hz                   │
│      Number of Harmonics: │9                            │
│      Output Variables:    │V(1)                         │
│                                                        │
│   ☐ Include detailed bias point information for nonlinear │
│      controlled sources and semiconductors (/OP)        │
│                                                        │
└────────────────────────────────────────────────────────┘
```

To start the simulation:

1.   Click on **PSpice**, click on **Run** and click on **Trace.**
2.   Select V(1); it will appear as a function of time.
3.   To obtain its **Fourier spectrum**, in the **Trace** box, click on **Fourier.** The spectrum of V(1) will appear on the **PROBE** screen. It is shown next.

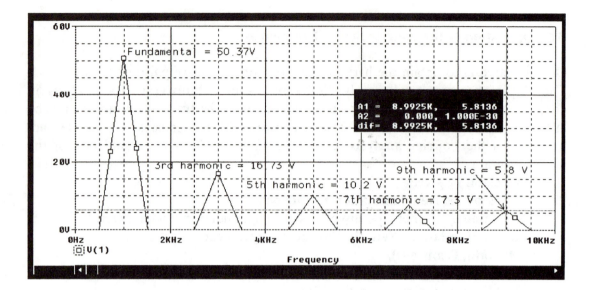

Each of the peaks present represents the amplitude and the frequency of a specific harmonic. For example, the **fundamental frequency**, or the first harmonic, is present at its frequency of 1 kHz, having an amplitude of 50.37 V. We note that only odd harmonics of diminishing amplitudes are present. It was this information that was used in Figure 10.02 to set the voltages and the frequencies of the five harmonic voltage sources.

We can also get a numerical printout of the harmonics from the **Output File**. To obtain it, in **PSpice** dialog box, click on **View Output File.** To save space, only its relevant excerpt is shown next.

```
FOURIER COMPONENTS OF TRANSIENT RESPONSE V(1)

DC COMPONENT =  -3.960396E-01

HARMONIC   FREQUENCY    FOURIER    NORMALIZED    PHASE       NORMALIZED
   NO        (HZ)      COMPONENT   COMPONENT     (DEG)       PHASE (DEG)

    1      1.000E+03   5.093E+01   1.000E+00   -8.911E-01    0.000E+00
    2      2.000E+03   7.925E-01   1.556E-02   -9.178E+01   -9.000E+01
    3      3.000E+03   1.698E+01   3.334E-01   -2.673E+00    1.021E-14
    4      4.000E+03   7.936E-01   1.558E-02   -9.356E+01   -9.000E+01
    5      5.000E+03   1.020E+01   2.002E-01   -4.455E+00    7.994E-15
    6      6.000E+03   7.955E-01   1.562E-02   -9.535E+01   -9.000E+01
    7      7.000E+03   7.290E+00   1.431E-01   -6.238E+00   -1.030E-13
    8      8.000E+03   7.983E-01   1.567E-02   -9.713E+01   -9.000E+01
    9      9.000E+03   5.677E+00   1.115E-01   -8.020E+00   -1.188E-12


TOTAL HARMONIC DISTORTION =   4.302880E+01 PERCENT

         JOB CONCLUDED

      TOTAL JOB TIME          .17
```

The **Output File** lists all nine harmonics, as we requested in the **Transient Output file Options** dialog box. The column headed **FOURIER COMPONENT** lists the amplitude of each of them. The even harmonics are of very small amplitude compared to the odd harmonics. This is brought out visually by their **PROBE** traces. Only the odd harmonics are shown in the **PROBE** plot.

The next column, **NORMALIZED COMPONENT,** is the ratio of the amplitude of any harmonics with the **NO 1** harmonic. For example, the third harmonic has a normalized component of 16.98/50.93 = .33. It again shows how much smaller the even harmonics are compared to the odd ones. For example, the fourth harmonic has a normalized component of only .01558.

The **PHASE** column shows the relative phase between a particular harmonic and the fundamental frequency. It is shown that the even harmonics all have -90° phase shifts. Furthermore, as we noted, their amplitudes are negligible compared to those of the odd harmonics. In consequence, the **Fourier series** representing the square pulse of Figure 10.02 consists of sine terms only. **The NORMALIZED PHASE (DEG)** is the phase of any harmonic relative to the fundamental.

Using the preceding data from the **Output File** and the **PROBE** plots allows us to write the mathematical expression for the **Fourier series** of our square pulse.

$$V(1) = 0 + 50.93\ sin\omega_0t + 16.98\ sin3\omega_0t + 10.2\ sin5\omega_0t + 7.3\ sin7\omega_0t + 5.7\ sin9\omega_0t\ \ V$$

The **DC COMPONENT** of the square pulse is listed as -.39 volts, thus in the above summation, it was neglected. We can tell by inspection that the dc value of the square pulse is zero volts.

Total Harmonic Distortion

The listing in the Output File showed the **total harmonic distortion** to be equal to 42.89%. It is a measure of the deviation of a composite waveform from the waveform of its fundamental harmonic. In our case, that difference was between a square wave and a sine wave. For this case, harmonic distortions of about 40% are the norm. As a signal undergoes transmission from source to load through an electrical network, a change of its harmonic distortion would indicate a change in the shape of the output signal compared to the input signal. The smaller that change, the more nearly the output signal resembles the input signal. **Harmonic distortion** is usually expressed as a percentage. For each harmonic, the percent harmonic distortion is defined as

$$\text{Percent distortion} = \frac{V_n}{V_1} * 100$$

The **total harmonic distortion (THD)** involving all the harmonics is given as:

$$\% \text{ total distortion} = [(\% \text{ second})^2 + (\% \text{ third})^2 + \ldots + (\% \text{ nth})^2]^{1/2}$$

Let us verify the **THD** calculated by the **PSpice** program by summing the percent harmonic distortions contributed by each of them. The results are tabulated next.

Harmonic Number	Percent Distortion	(Percent Distortion)^2
2	.314	.098
3	33.3	1110.0
4	.314	.098
5	20.0	400.0
6	.314	.098
7	14.3	204.0
8	.314	.098
9	1.1	1.23

We now form the sum of the contents of the last column to obtain:

$$(.098 + 1110 + .098 + 400 + .098 + 204 + .098 + 1.23)^{1/2} = 41.4\%$$

This is in reasonable agreement with the result calculated by the **PSpice** program as shown on page 332. The even harmonics produced the smallest percent distortions because of their small amplitudes. Conversely, the odd harmonics, because of their large amplitudes, produced the largest percent distortions.

The RMS Value of a Fourier Series

We recall that the real power delivered by a current or voltage source is directly proportional to the **RMS** value of a current or voltage signal. A nonsinusoidal waveform, such as the square wave under investigation, can be represented by a sum of sine waves. Therefore, it becomes apparent that if we can obtain the effective **RMS** value of that sum, we can than calculate the power delivered to a load by a nonsinusoidal current or voltage source.

To obtain the effective **RMS** value of a nonsinusoidal signal, we need to sum the squares of the **RMS** value, designated as E_n, of each harmonic and then take the square root of that sum. In symbols:

$$E_{(Fourier\ series)} = (E_1{}^2 + E_2{}^2 + E_3{}^2 + \ .\ .\ .\ .\ E_n{}^2)^{1/2}\ volts$$

We shall calculate the effective **RMS** value of the square pulse voltage **V1** of Figure 10.01 from the data of the **FOURIER COMPONENTS OF TRANSIENT RESPONSE V(1)** contained in the **Output File** listed on page 332. Note carefully that the peak value of the harmonics are listed. To convert them to **RMS** values, multiply each by .707. The results are tabulated next.

Harmonic Number	E	E^2
1	36.0	1299.1
2	.1	.01
3	12.0	144.0
4	.1	.01
5	7.2	51.8
6	.1	.01
7	5.1	30.3
8	.1	.01
9	4.0	16.0
		Sum: 1541.2

Taking the roots of the sum of the squares of E^2, we get

$$E_{(Fourier\ series)} = (1541.2)^{1/2} = 39.3\ V_{(RMS)}$$

This is close to the anticipated value of 40 $V_{(RMS)}$ for the square wave of the voltage source **V1**, or its nodal equivalent V(1), in Figure 10.01. A point well worth remembering is that the **RMS** value of a square wave is always equal to its amplitude.

The Power Delivered by a Fourier Series

Let us obtain the total **RMS** power delivered to **R1** by voltage source **V1** shown in Figure 10.03.

Figure 10.03

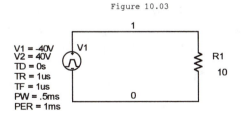

V1 = -40V
V2 = 40V
TD = 0s
TR = 1us
TF = 1us
PW = .5ms
PER = 1ms

To start, we enlist the help of **PROBE** and get the trace of the **RMS** power to **R1**. The result is shown below. The value of that power is 159.9 watts. If we compute the **RMS** power from **RMS** value of the **Fourier series,** 39.3 V$_{(RMS)}$, which we obtained earlier, we get

$$P(R1) = \frac{(39.3)^2}{10} = 154.4 watts$$

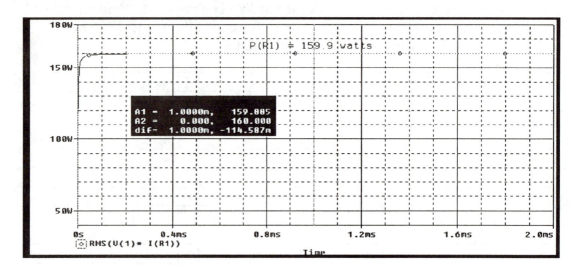

The two values are in good agreement. We shall next compute the contributions, both numerical and percent, that each harmonic makes to the power to **R1**. In the compilation of these contributions, we neglect those made by the even harmonics due to their smallness. The results are compiled next.

Harmonic Number	P$_{nth}$ Harmonic	% P$_{nth}$ Harmonic
1	129.9 W	81.0%
3	14.4 W	9.0%
5	5.2 W	3.2%
7	3.0 W	1.9%
9	1.6 W	1.0%
Sum	154.1 W	96.1%

The total power delivered to **R1** by the odd harmonics is about even with the previously obtained value for that power. In general, the fundamental frequency, or the first harmonic, will deliver the largest percentage of power to a load. The higher harmonics make ever smaller contributions to that power.

The **Fourier theorem** is a general theorem that applies to many different physical systems. For example, if sport coaches did not eliminate the fundamental frequency from their voices when they were shouting at their players, their voices would be far more audible, less high pitched and more durable.

FOURIER ANALYSIS OF A RESISTIVE VOLTAGE-DIVIDER CIRCUIT

The circuit shown in Figure 10.04 is a resistive voltage divider. It will demonstrate that it does alter the amplitudes of all harmonics proportionately so that the shapes of the output voltage V(2) and the input voltage V(1) are the same. Their only difference is one of scale.

Figure 10.04

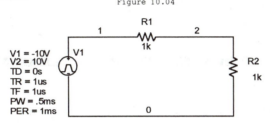

The **PROBE** plot shows the voltages V(1) and V(2) as a function of time. We performed a **Transient analysis** of 2 milliseconds to obtain these results. The only difference between these two voltage is that the amplitude of the output voltage V(2) is one-half the amplitude of the input voltage V(1). Both, however, are square waves.

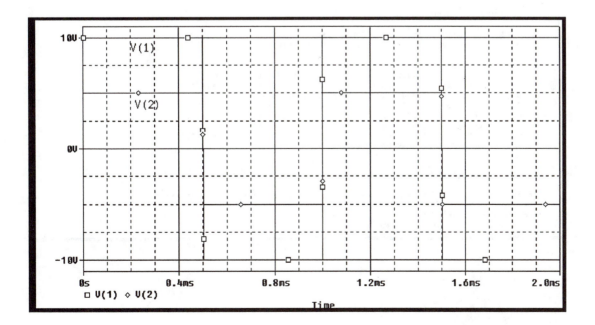

Next, a **Fourier analysis** was performed, with the results shown.

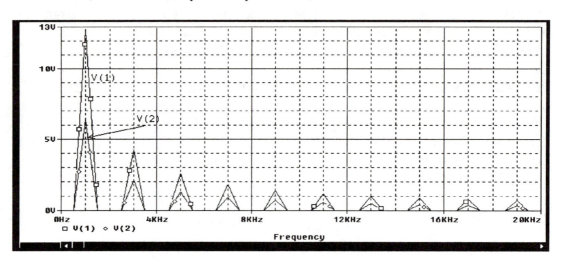

The **PROBE** data shows that the two spectra are scale replicas of each other. Each harmonic of V(2) is one half the value of its corresponding harmonic of V(1). Thus, there is no distortion between the two voltages. Both are square waves. Let us turn to the data contained in the **Output File**. That data shown corroborates that each of the harmonics of V(2) is one half the amplitude to the corresponding harmonic of V(1). Also, both voltages V(1) and V(2) have the same 43% harmonic distortion. Hence, their shapes are identical.

```
FOURIER COMPONENTS OF TRANSIENT RESPONSE V(1)

 DC COMPONENT =  -9.900990E-02

 HARMONIC    FREQUENCY      FOURIER      NORMALIZED      PHASE        NORMALIZED
   NO          (HZ)        COMPONENT     COMPONENT       (DEG)        PHASE (DEG)

    1        1.000E+03     1.273E+01     1.000E+00     -8.911E-01     0.000E+00
    2        2.000E+03     1.981E-01     1.556E-02     -9.178E+01     -9.000E+01
    3        3.000E+03     4.246E+00     3.334E-01     -2.673E+00     1.021E-14
    4        4.000E+03     1.984E-01     1.558E-02     -9.356E+01     -9.000E+01
    5        5.000E+03     2.549E+00     2.002E-01     -4.455E+00     7.994E-15
    6        6.002E+03     1.989E-01     1.562E-02     -9.535E+01     -9.000E+01
    7        7.000E+03     1.823E+00     1.431E-01     -6.238E+00     -1.030E-13
    8        8.000E+03     1.996E-01     1.567E-02     -9.713E+01     -9.000E+01
    9        9.000E+03     1.419E+00     1.115E-01     -8.020E+00     -1.188E-12


    TOTAL HARMONIC DISTORTION =    4.302880E+01 PERCENT

**** 05/04/00 11:29:30 *********** Evaluation PSpice (Mar 1999) **************

 ** circuit file for profile: 1
```

```
FOURIER COMPONENTS OF TRANSIENT RESPONSE V(2)

DC COMPONENT =  -4.950495E-02

HARMONIC    FREQUENCY     FOURIER     NORMALIZED      PHASE       NORMALIZED
  NO          (HZ)       COMPONENT    COMPONENT       (DEG)      PHASE (DEG)

   1        1.000E+03    6.366E+00    1.000E+00     -8.911E-01    0.000E+00
   2        2.000E+03    9.906E-02    1.556E-02     -9.178E+01   -9.000E+01
   3        3.000E+03    2.123E+00    3.334E-01     -2.673E+00    1.021E-14
   4        4.000E+03    9.920E-02    1.558E-02     -9.356E+01   -9.000E+01
   5        5.000E+03    1.275E+00    2.002E-01     -4.455E+00    7.994E-15
   6        6.000E+03    9.944E-02    1.562E-02     -9.535E+01   -9.000E+01
   7        7.000E+03    9.113E-01    1.431E-01     -6.238E+00   -1.030E-13
   8        8.000E+03    9.978E-02    1.567E-02     -9.713E+01   -9.000E+01
   9        9.000E+03    7.097E-01    1.115E-01     -8.020E+00   -1.188E-12


   TOTAL HARMONIC DISTORTION =   4.302880E+01 PERCENT
```

Square Wave with DC Voltage

We shall shift the voltage **V1** upward to give it an average voltage of 10 volts. Our objective is to find the **Fourier spectra** of V(1) and V(2). The circuit used is shown in Figure 10.05.

Figure 10.05

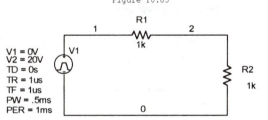

The result of a 2 millisecond **transient analysis** is shown next. Again, we observe that the output voltage V(2) is a scale replica of the input voltage V(1). No distortion of the voltage waveforms is in evidence.

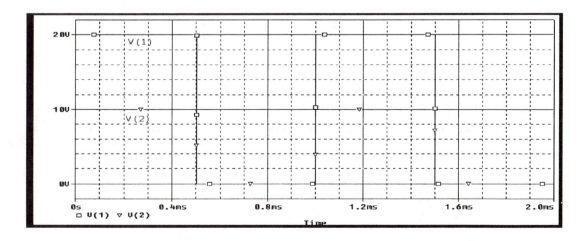

Shown next is the **Fourier spectra** of these two voltages. The most obvious difference between these two traces of V(1) and V(2) compared to the traces of these voltages in Figure 10.04 is that now both show a dc component. The dc voltage of V(1) is 10 volts because of the shift of V(1) and the dc voltage of V(2) is one half of that dc voltage, 5 volts, due to the action of the voltage divider.

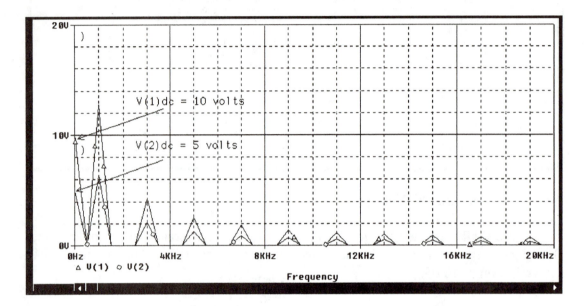

We need to ask what happens, if anything, to the values of the harmonics present. To get the answer, we consult the **Output file**.

```
*************************************************************************

FOURIER COMPONENTS OF TRANSIENT RESPONSE V(1)

  DC COMPONENT =   9.900990E+00

  HARMONIC   FREQUENCY    FOURIER     NORMALIZED    PHASE        NORMALIZED
    NO         (HZ)      COMPONENT    COMPONENT     (DEG)       PHASE (DEG)

     1       1.000E+03    1.273E+01   1.000E+00    -8.911E-01    0.000E+00
     2       2.000E+03    1.981E-01   1.556E-02    -9.178E+01   -9.000E+01
     3       3.000E+03    4.246E+00   3.334E-01    -2.673E+00   -2.551E-10
     4       4.000E+03    1.984E-01   1.558E-02    -9.356E+01   -9.000E+01
     5       5.000E+03    2.549E+00   2.002E-01    -4.455E+00   -1.266E-09
     6       6.000E+03    1.989E-01   1.562E-02    -9.535E+01   -9.000E+01
     7       7.000E+03    1.823E+00   1.431E-01    -6.238E+00   -3.504E-09
     8       8.000E+03    1.996E-01   1.567E-02    -9.713E+01   -9.000E+01
     9       9.000E+03    1.419E+00   1.115E-01    -8.020E+00   -7.393E-09


    TOTAL HARMONIC DISTORTION =   4.302880E+01 PERCENT
```

```
    DC COMPONENT =     4.950495E+00

    HARMONIC    FREQUENCY     FOURIER    NORMALIZED     PHASE       NORMALIZED
       NO         (HZ)       COMPONENT   COMPONENT      (DEG)      PHASE (DEG)

        1       1.000E+03    6.366E+00   1.000E+00    -8.911E-01    0.000E+00
        2       2.000E+03    9.906E-02   1.556E-02    -9.178E+01   -9.000E+01
        3       3.000E+03    2.123E+00   3.334E-01    -2.673E+00   -2.551E-10
        4       4.000E+03    9.920E-02   1.558E-02    -9.356E+01   -9.000E+01
        5       5.000E+03    1.275E+00   2.002E-01    -4.455E+00   -1.266E-09
        6       6.000E+03    9.944E-02   1.562E-02    -9.535E+01   -9.000E+01
        7       7.000E+03    9.113E-01   1.431E-01    -6.238E+00   -3.504E-09
        8       8.000E+03    9.978E-02   1.567E-02    -9.713E+01   -9.000E+01
        9       9.000E+03    7.097E-01   1.115E-01    -8.020E+00   -7.393E-09


    TOTAL HARMONIC DISTORTION =     4.302880E+01 PERCENT
```

The data shows that neither the harmonics of V(1) and V(2), nor the **harmonic distortion** of 43% change in comparison to the harmonics and their distortion in the circuit of Figure 10.04. The only difference is in the new dc values of 10 volts for V(1) and 5 volts for V(2). This corresponds with the **PROBE** traces of these two voltages.

Shifting the Input Voltage Pulse V1 in Time

We next delay the voltage pulse **V1** by a **TD** of .5 milliseconds. Our objective is to find the effects of that shift on the harmonic spectra of V(1) and V(2). We use the circuit of Figure 10.06. It is essentially the same circuit as in Figure 10.05.

Figure 10.06

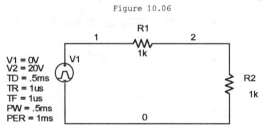

The effect of the **TD** of .5 milliseconds is to shift the voltage pulse **V1** and the voltage V(2) to the right on the **X-axis**. A negative **TD** would have moved the pulse **V1** and V(2) to the left on the **X-axis**.

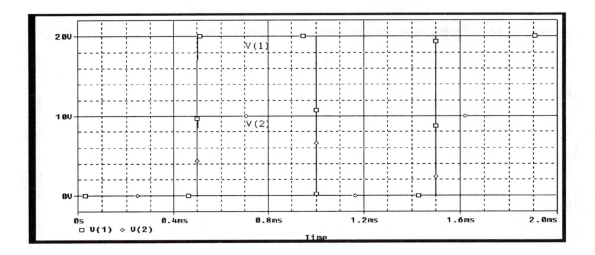

What of the harmonic spectrum as shown on the **PROBE** pot? There is no change from the previous plot! But there have got to be some changes.

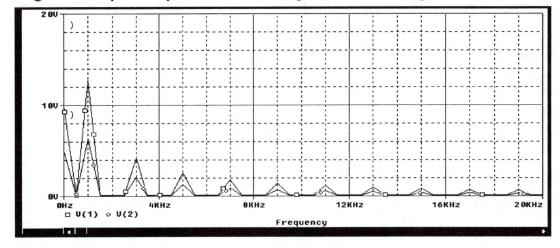

Let us look at the data from the **Output File** shown on next page. The data from the **Output File** shows that neither the dc values nor the amplitudes of any harmonics were changed from their previous values. However, the phase of every odd harmonics has been shifted by 180°. A **TD** of .5 milliseconds corresponds to a delay of one-half the period of the voltage pulse **V1** and consequently to a phase shift of 180°.

```
FOURIER COMPONENTS OF TRANSIENT RESPONSE V(1)

  DC COMPONENT =    1.029704E+01

  HARMONIC    FREQUENCY      FOURIER     NORMALIZED      PHASE        NORMALIZED
    NO          (HZ)        COMPONENT    COMPONENT       (DEG)        PHASE (DEG)

      1       1.000E+03     1.275E+01    1.000E+00      1.773E+02     0.000E+00
      2       2.000E+03     5.941E-01    4.661E-02      8.941E+01    -2.653E+02
      3       3.000E+03     4.282E+00    3.360E-01      1.720E+02    -3.600E+02
      4       4.000E+03     5.942E-01    4.662E-02      8.881E+01    -6.205E+02
      5       5.000E+03     2.610E+00    2.048E-01      1.668E+02    -7.198E+02
      6       6.000E+03     5.944E-01    4.663E-02      8.821E+01    -9.758E+02
      7       7.000E+03     1.907E+00    1.496E-01      1.618E+02    -1.079E+03
      8       8.000E+03     5.946E-01    4.665E-02      8.761E+01    -1.331E+03
      9       9.000E+03     1.526E+00    1.197E-01      1.571E+02    -1.439E+03


     TOTAL HARMONIC DISTORTION =    4.474822E+01 PERCENT
```

```
FOURIER COMPONENTS OF TRANSIENT RESPONSE V(2)

   DC COMPONENT =    5.148520E+00

  HARMONIC    FREQUENCY      FOURIER     NORMALIZED      PHASE        NORMALIZED
    NO          (HZ)        COMPONENT    COMPONENT       (DEG)        PHASE (DEG)

      1       1.000E+03     6.373E+00    1.000E+00      1.773E+02     0.000E+00
      2       2.000E+03     2.971E-01    4.661E-02      8.941E+01    -2.653E+02
      3       3.000E+03     2.141E+00    3.360E-01      1.720E+02    -3.600E+02
      4       4.000E+03     2.971E-01    4.662E-02      8.881E+01    -6.205E+02
      5       5.000E+03     1.305E+00    2.048E-01      1.668E+02    -7.198E+02
      6       6.000E+03     2.972E-01    4.663E-02      8.821E+01    -9.758E+02
      7       7.000E+03     9.533E-01    1.496E-01      1.618E+02    -1.079E+03
      8       8.000E+03     2.973E-01    4.665E-02      8.761E+01    -1.331E+03
      9       9.000E+03     7.629E-01    1.197E-01      1.571E+02    -1.439E+03


     TOTAL HARMONIC DISTORTION =    4.474822E+01 PERCENT
```

FOURIER SERIES OF COMMON TIME FUNCTIONS

The Triangular, or Saw Tooth, Wave

This section covers some common time functions and their **Fourier transforms** that are often found in electronic circuits. We start with the triangular, or sawtooth, wave. Such a wave is used to sweep an electron beam across an oscilloscope screen. Figure 10.07 applies such a wave across our previously employed resistive voltage divider.

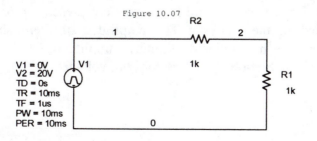

Figure 10.07

A **transient analysis** of 40 milliseconds produced the traces of **the time functions** of the voltages of V(1) and V(2). Carefully note the specified parameters for the voltage pulse V1. The pulse width **(PW)** and the period **(PER)** are both equal to 10 milliseconds.

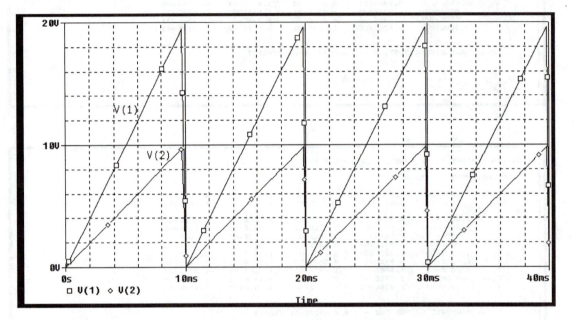

The peak value of V(2) is determined by the voltage of V(1) and the voltage-divider ratio of our circuit. Thus, that peak is 5 volts. No distortion of V(2) relative to V(1) has occurred; thus, we can anticipate that both V(1) and V(2) will have the same harmonic distortion.

To perform a **Fourier analysis**, we need to specify the fundamental frequency of that pulse.   Since the period of the pulse is 10 milliseconds, therefore, the fundamental frequency of the pulse specified to run the Fourier analysis is 100 Hz. The reader is referred back to the point made above that the fundamental frequency determines the period of a pulse.

· The data produced by the **Fourier analysis** of this circuit is shown next.
The spectrum of V(1) shows a dc voltage of 10 volts and that of V(2) shows a dc voltage of 5 volts. Both even and odd harmonics are present up to the limit of nine harmonics specified in the analysis. The amplitude of them shows the general tendency of higher order harmonics declining in amplitude. Of note is the relatively large **total harmonic distortion** of 73.8 % for both voltages.

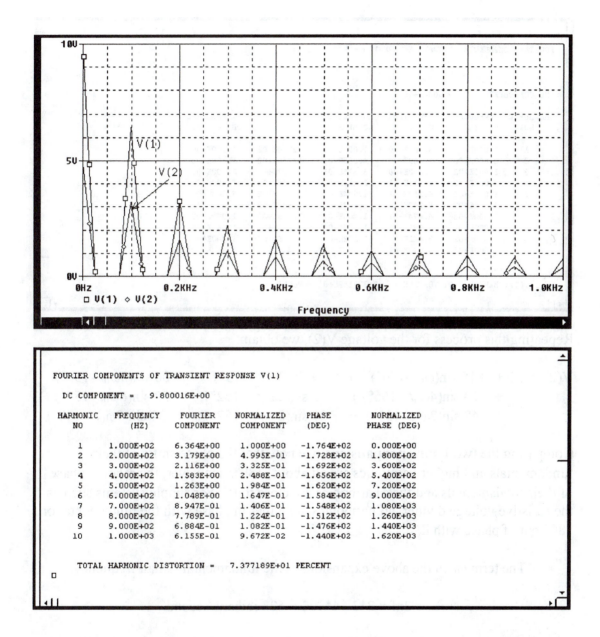

The data from the **Output File** for V(1) shown above and that for V(2) shown below allows us to write the **Fourier series** . For the voltage V(1) we have

$$V(1) = 9.8 + 6.36 \sin(\omega_0 t - 176°) + 3.179 \sin(2\omega_0 t - 173°) + 2.11 \sin(3\omega_0 t - 169°)$$
$$+ 1.58 \sin(4\omega_0 t - 166°) + 1.26 \sin(5\omega_0 t - 162°) + 1.05 \sin(6\omega_0 t - 158°)$$
$$+ .89 \sin(7\omega_0 t - 155°) + .78 \sin(8\omega_0 t - 152°) + .69 \sin(9\omega_0 t - 148°)$$

```
        FOURIER COMPONENTS OF TRANSIENT RESPONSE V(2)


        DC COMPONENT =    4.900008E+00

        HARMONIC   FREQUENCY    FOURIER    NORMALIZED    PHASE      NORMALIZED
           NO        (HZ)      COMPONENT   COMPONENT     (DEG)     PHASE (DEG)

            1      1.000E+02   3.182E+00   1.000E+00   -1.764E+02   0.000E+00
            2      2.000E+02   1.589E+00   4.995E-01   -1.728E+02   1.800E+02
            3      3.000E+02   1.058E+00   3.325E-01   -1.692E+02   3.600E+02
            4      4.000E+02   7.916E-01   2.488E-01   -1.656E+02   5.400E+02
            5      5.000E+02   6.314E-01   1.984E-01   -1.620E+02   7.200E+02
            6      6.000E+02   5.242E-01   1.647E-01   -1.584E+02   9.000E+02
            7      7.000E+02   4.474E-01   1.406E-01   -1.548E+02   1.080E+03
            8      8.000E+02   3.895E-01   1.224E-01   -1.512E+02   1.260E+03
            9      9.000E+02   3.442E-01   1.082E-01   -1.476E+02   1.440E+03
           10      1.000E+03   3.078E-01   9.672E-02   -1.440E+02   1.620E+03


        TOTAL HARMONIC DISTORTION =   7.377189E+01 PERCENT
```

Repeating this process for the voltage V(2), we obtain:

$$V(2) = 4.9 + 3.18 \sin(\omega_0 t - 176°) + 1.59 \sin(2\omega_0 t - 173°) + 1.06 \sin(3\omega_0 t - 169°)$$
$$+ .79 \sin(4\omega_0 t - 166°) + .63 \sin(5\omega_0 t - 162°) + .52 \sin(6\omega_0 t - 158°)$$
$$+ .48 \sin(7\omega_0 t - 155°) + .39 \sin(8\omega_0 t - 152°) + .31 \sin(9\omega_0 t - 148°)$$

Comparing the two **Fourier expansions**, we note that the phase shifts of their fundamentals and higher harmonics are all about -180°. However, the relative phase of their fundamentals and their harmonics are close to 0°. For simple systems such as the resistive voltage divider, all harmonics are either in phase with the fundamental or 180° out of phase with it.

The term ω_0 in the above expansions is the fundamental radian frequency:

$$\omega_0 = 2\pi f = 2*3.14*100 = 628 \text{ radians/second}$$

If we consult a standard reference book to obtain the amplitude of the **Fourier coefficients**, the following formula is given:

$$V_n = \frac{Vm}{3.14n}$$

where : n = number of harmonics of interest

V_n = the amplitude of the nth harmonic

Vm = the maximum amplitude of triangular wave: 10 volts for V(1), 5 volts for V(2)

The Half Wave Rectified Sine Wave

The half-wave rectified wave is very common and is the output voltage of a rectifier circuit. The objective of such a circuit is to derive a dc voltage from an input voltage such as a sine voltage, which does not have a dc component. An important parameter of circuit performance is the amount of **ripple** present in the output voltage. Ripple is an index of the variation of the output voltage in addition to its dc component. It is related to the harmonic content of the output voltage. If no harmonics were present in that voltage, the ripple would be zero and the output voltage would be a pure dc signal This is striven for but never perfectly obtained.

We shall use the circuit in Figure 10.08 to study the effect of rectifier operation.

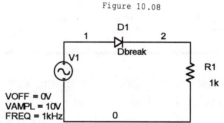

Figure 10.08

This circuit contains a diode. It is a semiconductor that passes current in the direction of the arrowhead, ideally offering no resistance to the flow of current in that, the forward, direction. It opposes the flow of current against the direction of the arrowhead, ideally offering infinite resistance to the flow of current in that, the reverse direction.

To get the diode, proceed as follows:
1. In the **Place Part** dialog box, click on **Add Library. . .**, the **Browse File** dialog box opens.
2. Click on **breakout**, click on **OPEN.**
3. Scroll to **Dbreak,** click on **OK.**
4. Place the diode where desired.

During the positive portion of **V1**, current will flow, developing a voltage V(2) that is a close replica of **V1**. During the negative portion of **V1**, no current will flow and the voltage V(2) will be zero volts. A **transient analysis** of 2 milliseconds duration will be performed together with a **Fourier analysis**. The 1 kHz frequency of **V1** is the center frequency for that analysis.

We start our analysis with the data provided by the **Transient analysis**.

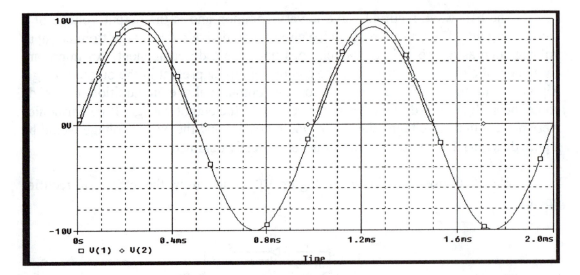

There are no surprises based upon our previous comments. Note, however, that the peak voltage of V(2) is slightly less that that of V(1). This is because of the small, yet finite, forward resistance of the non-ideal diode used in our simulation. For many diodes, a voltage drop of .5 volt to .7 volts is the norm. Next, we turn to the **Fourier data** shown.

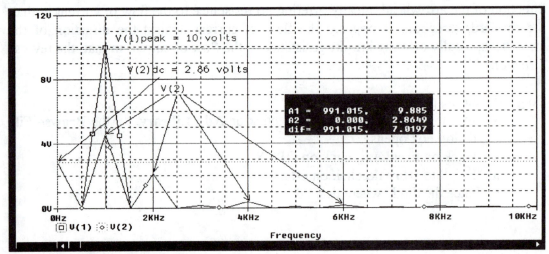

The fundamental frequency of V(1) is represented by only one harmonic, the fundamental at 1 kHz. Its amplitude is 10 volts. The spectra of V(2) shows a dc value of about 2.86 volts dc. Theoretically, the dc component of a half-wave rectified sine wave is

$$V(2)_{dc} = \frac{V(1)_{max}}{3.14} = \frac{10}{3.14} = 3.18V$$

The difference between the ideal calculated and the actual measured voltage of V(2) on the **PROBE** plot is accounted for by the voltage drop across the nonideal diode.

The spectra of V(2) consists of a relatively large fundamental harmonic and the usual diminishing higher harmonics. The even harmonics are relatively larger than the odd harmonics. In most textbooks of the harmonic spectrum of V(2), the odd harmonics are usually neglected due to their small size.

We turn next to the **Output file** data.

```
FOURIER COMPONENTS OF TRANSIENT RESPONSE V(1)


DC COMPONENT =    2.460231E-04

HARMONIC   FREQUENCY    FOURIER    NORMALIZED    PHASE       NORMALIZED
   NO        (HZ)      COMPONENT   COMPONENT     (DEG)      PHASE (DEG)

   1       1.000E+03   9.987E+00   1.000E+00   9.277E-05    0.000E+00
   2       2.000E+03   8.171E-04   8.182E-05   7.672E+01    7.672E+01
   3       3.000E+03   1.228E-03   1.230E-04   7.117E+01    7.117E+01
   4       4.000E+03   6.713E-04   6.722E-05   7.272E+01    7.272E+01
   5       5.000E+03   8.344E-04   8.355E-05   6.800E+01    6.800E+01
   6       6.000E+03   5.697E-04   5.705E-05   6.789E+01    6.789E+01
   7       7.000E+03   6.523E-04   6.532E-05   6.334E+01    6.334E+01
   8       8.000E+03   4.759E-04   4.765E-05   6.831E+01    6.831E+01
   9       9.000E+03   5.224E-04   5.231E-05   6.349E+01    6.349E+01
  10       1.000E+04   4.196E-04   4.202E-05   7.534E+01    7.534E+01


   TOTAL HARMONIC DISTORTION =   2.182081E-02 PERCENT
```

In the spectra of V(1), only the fundamental harmonic is present in strength. All other harmonics are essentially zero volts. Since there are no distorting harmonics present, the **total harmonic distortion** is only .0285%.

```
FOURIER COMPONENTS OF TRANSIENT RESPONSE V(2)


DC COMPONENT =   2.834341E+00

HARMONIC   FREQUENCY    FOURIER     NORMALIZED    PHASE       NORMALIZED
   NO        (HZ)      COMPONENT    COMPONENT    (DEG)       PHASE (DEG)

    1       1.000E+03   4.545E+00   1.000E+00   -2.145E-01    0.000E+00
    2       2.000E+03   2.100E+00   4.621E-01   -9.002E+01   -8.959E+01
    3       3.000E+03   1.403E-01   3.087E-02   -1.733E+02   -1.726E+02
    4       4.000E+03   4.127E-01   9.079E-02   -9.011E+01   -8.925E+01
    5       5.000E+03   8.161E-02   1.796E-02   -1.683E+02   -1.672E+02
    6       6.000E+03   1.735E-01   3.816E-02   -9.040E+01   -8.912E+01
    7       7.000E+03   5.705E-02   1.255E-02   -1.631E+02   -1.616E+02
    8       8.000E+03   9.453E-02   2.080E-02   -9.103E+01   -8.932E+01
    9       9.000E+03   4.373E-02   9.622E-03   -1.579E+02   -1.560E+02
   10       1.000E+04   5.910E-02   1.300E-02   -9.212E+01   -8.998E+01


   TOTAL HARMONIC DISTORTION =   4.746830E+01 PERCENT
```

The spectrum of V(2) confirms the **PROBE** data above. The dc component of V(2) is listed as 2.83 volts. The first harmonic at 4.55 volts is the strongest by far. The second harmonic is already down to 2 volts. All higher harmonics are decaying rapidly. We now write the **Fourier expansion** of V(2).

$$V(2) = 2.83 + 4.55 \sin(\omega_0 t - .215°) + 2.10 \sin(2\omega_0 t - 90°) + .14 \sin(3\omega_0 t - 177°)$$
$$.41 \sin(4\omega_0 t - 90.8°) + .08 \sin(5\omega_0 t - 168°) + .17 \sin(6\omega_0 t - 90.4°)$$
$$.05 \sin(7\omega_0 t - 163°) + .09 \sin(8\omega_0 t - 91°) + .04 \sin(9\omega_0 t - 156°)$$

The Fully Rectified Sine wave

The halfwave rectifier produced a 2.83 volt dc wave from a 10 volt sine voltage. Should we desire to increase the desired dc voltage, for the price of three more diodes we can obtain a higher dc voltage from a sine wave. The circuit to do this is the popular full-wave rectifier, shown in Figure 10.9.

Figure 10.09

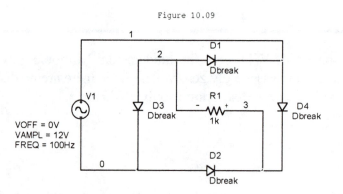

The output voltage is taken across resistor R1. Polarity markers have been placed near that resistor to identify its positive and negative terminals. A **transient analysis** of 20 milliseconds produced the traces of the input voltage V(1) and the output voltage V(3,2). The trace of V(3,2) is that of a fully rectified sine wave. The peak voltage of V(3,2) is 1.44 volts less than the peak voltage of V(1) because two conducting diodes in series produce a total voltage drop of about 1.4 volts.

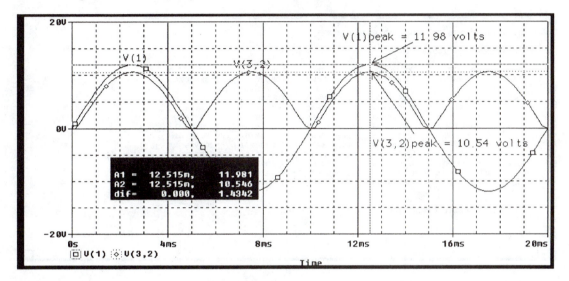

Next shown are the **Fourier spectra** of V(1) and V(3,2). Two plots were used to make the difference in the two spectra more obvious. The nodal voltage V(1), which is equal to the source voltage **V1**, has an amplitude of 12 V and a frequency of 100 Hz. There are no harmonics other than the fundamental one present because V(1) is a pure sine wave.

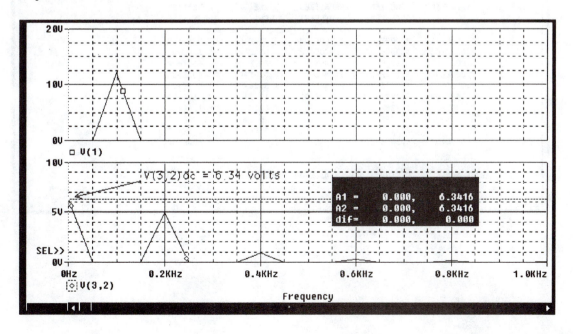

The **Fourier spectrum** of V(3,2) has a dc value of 6.34 volts, up from 2.83 volts for the half-wave rectifier. The output voltage V(3,2) does not contain a fundamental harmonic, only even harmonics are present. Of them, the second harmonic has the largest amplitude. Higher harmonics quickly decay toward zero volts. The absence of a fundamental harmonic results in a **large total harmonic distortion,** as will be seen in the **Output File** data shown next.

```
FOURIER COMPONENTS OF TRANSIENT RESPONSE V(1)

DC COMPONENT =   1.358784E-03

HARMONIC   FREQUENCY    FOURIER     NORMALIZED    PHASE      NORMALIZED
   NO        (HZ)      COMPONENT    COMPONENT     (DEG)      PHASE (DEG)

    1      1.000E+02    1.198E+01    1.000E+00    1.975E-03    0.000E+00
    2      2.000E+02    8.762E-04    7.311E-05   -1.249E+02   -1.249E+02
    3      3.000E+02    7.642E-04    6.377E-05    1.087E+02    1.087E+02
    4      4.000E+02    6.372E-04    5.317E-05    7.222E+01    7.221E+01
    5      5.000E+02    3.962E-04    3.306E-05    7.812E+01    7.811E+01
    6      6.000E+02    5.114E-04    4.267E-05    9.080E+01    9.079E+01
    7      7.000E+02    5.405E-04    4.510E-05    7.974E+01    7.973E+01
    8      8.000E+02    4.767E-04    3.977E-05    7.620E+01    7.618E+01
    9      9.000E+02    4.576E-04    3.819E-05    8.248E+01    8.246E+01
   10      1.000E+03    5.093E-04    4.250E-05    7.968E+01    7.966E+01


   TOTAL HARMONIC DISTORTION =   1.484355E-02 PERCENT
```

```
FOURIER COMPONENTS OF TRANSIENT RESPONSE V(3,2)


DC COMPONENT =   6.277224E+00

HARMONIC   FREQUENCY    FOURIER     NORMALIZED    PHASE      NORMALIZED
   NO        (HZ)      COMPONENT    COMPONENT     (DEG)      PHASE (DEG)

    1      1.000E+02    3.002E-02    1.000E+00   -8.770E+01    0.000E+00
    2      2.000E+02    4.948E+00    1.648E+02   -9.003E+01    8.536E+01
    3      3.000E+02    2.926E-02    9.746E-01   -9.506E+01    1.680E+02
    4      4.000E+02    9.139E-01    3.044E+01   -9.028E+01    2.605E+02
    5      5.000E+02    2.957E-02    9.848E-01   -9.652E+01    3.420E+02
    6      6.000E+02    3.489E-01    1.162E+01   -9.113E+01    4.350E+02
    7      7.000E+02    2.928E-02    9.754E-01   -9.881E+01    5.151E+02
    8      8.000E+02    1.669E-01    5.560E+00   -9.303E+01    6.085E+02
    9      9.000E+02    2.913E-02    9.702E-01   -1.010E+02    6.883E+02
   10      1.000E+03    8.900E-02    2.965E+00   -9.692E+01    7.800E+02


   TOTAL HARMONIC DISTORTION =   1.681416E+04 PERCENT
```

This data for V(1) shows that its fundamental harmonic is the only one present. It dc component is near zero volts. The data for V(3,2) shows the absence of a fundamental and the presence of a strong second harmonic. The numerical data corroborates that obtained from the **PROBE** traces. We shall next write the **Fourier series** for V(3,2) from our data neglecting the odd harmonics.

$$V(3,2) = 6.28 + 4.94 \sin(2\omega_0 t - 90°) + .91 \sin(4\omega_0 t - 90.3°) + .35 \sin(6\omega_0 t - 91°)$$
$$+ .17 \sin(8\omega_0 t - 93°) + .09 \sin(10\omega_0 t - 97°) \text{ volts}$$

FOURIER ANALYSIS OF THE RC LOW-PASS FILTER

In previous chapters, we applied current and voltage pulses to circuits containing both resistive and reactive elements. The analysis performed was done in the **Time (transient) domain**. Now we shall perform a **Fourier analysis** of such circuits.

We apply the square voltage pulse **V1** to the **RC** low-pass filter circuit in Figure 10.10. The output voltage V(2) is taken across the capacitor.

Figure 10.10

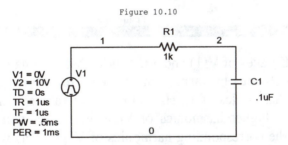

We begin our analysis by obtaining the voltages V(1) and V(2) as functions of time. We shall perform a 2 millisecond **transient analysis**.

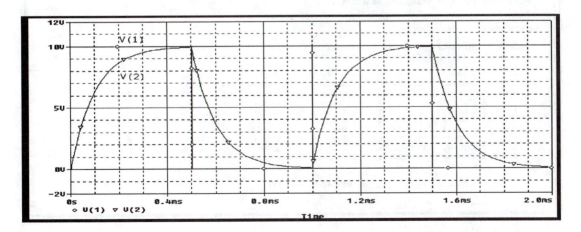

The **PROBE** traces show that the voltage V(2) charges up to the voltage of V(1) at a rate determined by the product **RC,** which is equal to .1 milliseconds. Thus, in .5 milliseconds, V(2) reaches 10 volts. This we have encountered before. Let us next obtain the **Fourier data** for this circuit. We start with the **PROBE** traces of V(1) and V(2).

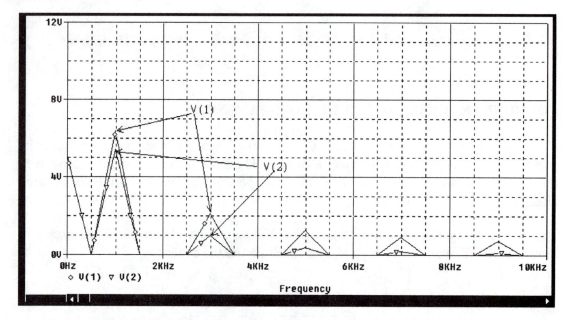

The **PROBE** traces of V(1) and V(2) show that both have the same dc voltage of 5 volts. This is obvious by inspection of their **Time domain** traces. Both have strong fundamental harmonics of 1 kHz and equal dc values. This is typical of a low pass filter. It is the higher harmonics of V(2) that suffer considerable attenuation when compared to the corresponding harmonics of V(1). This will be shown in their **Output File** data also.

```
FOURIER COMPONENTS OF TRANSIENT RESPONSE V(1)

DC COMPONENT =   4.950495E+00

HARMONIC    FREQUENCY      FOURIER     NORMALIZED     PHASE       NORMALIZED
   NO         (HZ)        COMPONENT    COMPONENT      (DEG)      PHASE (DEG)

    1       1.000E+03     6.366E+00    1.000E+00    -8.911E-01    0.000E+00
    2       2.000E+03     9.906E-02    1.556E-02    -9.178E+01   -9.000E+01
    3       3.000E+03     2.123E+00    3.334E-01    -2.673E+00   -2.551E-10
    4       4.000E+03     9.920E-02    1.558E-02    -9.356E+01   -9.000E+01
    5       5.000E+03     1.275E+00    2.002E-01    -4.455E+00   -1.266E-09
    6       6.000E+03     9.944E-02    1.562E-02    -9.535E+01   -9.000E+01
    7       7.000E+03     9.113E-01    1.431E-01    -6.238E+00   -3.504E-09
    8       8.000E+03     9.978E-02    1.567E-02    -9.713E+01   -9.000E+01
    9       9.000E+03     7.097E-01    1.115E-01    -8.020E+00   -7.393E-09
   10       1.000E+04     1.002E-01    1.574E-02    -9.891E+01   -9.000E+01


   TOTAL HARMONIC DISTORTION =   4.305759E+01 PERCENT
```

```
FOURIER COMPONENTS OF TRANSIENT RESPONSE V(2)

 DC COMPONENT =    5.008539E+00

 HARMONIC    FREQUENCY     FOURIER     NORMALIZED     PHASE        NORMALIZED
   NO          (HZ)       COMPONENT    COMPONENT      (DEG)       PHASE (DEG)

    1        1.000E+03    5.395E+00    1.000E+00    -3.254E+01    0.000E+00
    2        2.000E+03    1.100E-02    2.040E-03     3.281E+01    9.788E+01
    3        3.000E+03    9.956E-01    1.845E-01    -6.341E+01    3.420E+01
    4        4.000E+03    6.925E-03    1.284E-03     7.927E+00    1.381E+02
    5        5.000E+03    3.858E-01    7.151E-02    -7.460E+01    8.808E+01
    6        6.000E+03    5.121E-03    9.492E-04    -5.876E+00    1.893E+02
    7        7.000E+03    2.008E-01    3.722E-02    -8.026E+01    1.475E+02
    8        8.000E+03    4.221E-03    7.825E-04    -1.639E+01    2.439E+02
    9        9.000E+03    1.223E-01    2.267E-02    -8.374E+01    2.091E+02
   10        1.000E+04    3.708E-03    6.873E-04    -2.471E+01    3.007E+02


    TOTAL HARMONIC DISTORTION =    2.026796E+01 PERCENT
```

If we compare the **Fourier data** of V(1) with that of V(2), we can see the relative decline of the higher harmonics of V(2) compared to those of V(1). For example, the third harmonic component of V(1) is 2.12 volts, while that of V(2) is .995 volts. Their ratio is equal to .995volts/2.12 volts = .47. Let's find out why.

The impedance of the capacitor changes inversely with frequency. Thus, as the order of the harmonics increases, the impedance and reactance of the capacitor declines. Thus, the ratio V(2)/V(1) declines also. Let us tabulate the reactances of the capacitor and the ratio of V(2)/V(1) for the odd harmonics. The effect of the even harmonics is minor due to their small amplitudes and will be neglected.

Frequency (kHz)	Xc (Ω)	$V(2)/V(1) = -jXc/(R-jXc)$
1	1590	$.85\angle-32.1°$
3	531	$.47\angle-62.1°$
5	318	$.30\angle-72.4°$
7	227	$.22\angle-77.2°$
9	177	$.17\angle-80.0°$

We can now appreciate why the ratio of V(2)/V(1) calculated above for the third harmonic was .47. At that frequency, the ratio of the impedance of the capacitor to that of the circuit is equal to .47. In effect the circuit is a voltage divider. Also, we note that at 1 kHz, the ratio of V(2)/V(1) from the **PROBE** data, is 5.39 volts/6.36 volts = .85. That ratio corresponds to that in the calculated tabulation above. The reader is strongly encouraged to verify the voltage ratios for the remaining harmonics.

Rise Time of a Pulse and the Critical Frequency

The **rise time t_r** of a pulse is defined as the time is takes for a pulse to go from 10% of its initial value to 90% of its final value. Let us recall our **PROBE** plot of the trace of V(2) and determine its rise time.

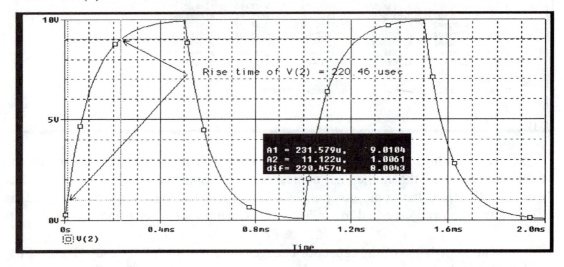

The rise time of V(2) measures 220.457 microseconds. We recall that the **critical frequency** of this circuit is calculated as

$$f_{(critical)} = 1/(2*314*R*C) \text{ Hz} = 1/(2*3.14*1 \text{ k}\Omega*.1 \text{ μF}) = 1.59 \text{ kHz}$$

The following relationship exists between the rise time of V(2) and the critical frequency of our circuit:

$$t_r = .35/f_c \text{ seconds}$$

Solving for f_c:

$$f_c = .35/(220.457 \text{ us}) = 1.59 \text{ kHz}$$

Thus, if we are in a laboratory and display the voltage V(2) on an oscilloscope, we are able to determine the **critical frequency** of this filter from a measurement of its rise time.

If we compare the rise time of V(1) of 1 microsecond with that of V(2), we notice the considerable increase in the rise time of V(2). The reason is this: it is the higher harmonics account for the rise time of the pulse. If they are eliminated, an increase in the rise time of the pulse is the result.

FOURIER ANALYSIS OF AN RC HIGH-PASS FILTER

By interchanging the positions of the resistor with that of the capacitor in Figure 10.10, we create a high-pass filter. Our objective again is to perform a **Fourier series analysis** in conjunction with a **transient analysis**. We use the circuit in Figure 10.11. Its component values and its input voltage V(1) are the same as for Figure 10.10.

Figure 10.11

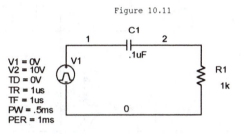

A **transient analysis** of 2 milliseconds produced the traces of V(1) and V(2) shown.

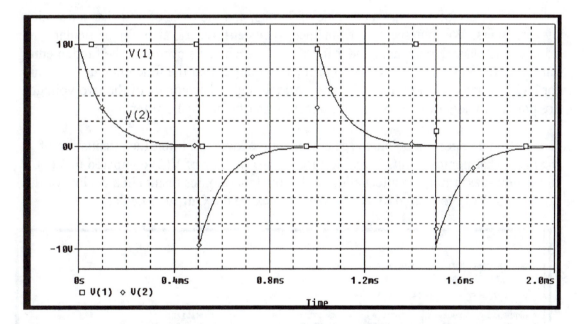

This circuit is a high-pass filter. Thus, we can, and will confirm that the lower frequencies suffer attenuation. The rise time of V(2) is identical to that of V(1). This is because the high frequencies are not eliminated as in the low-pass filter circuit. However, there seems to be a considerable fall time, t_f. We shall determine it below. The output voltage is taken across **R1**. As V(1) changes between zero volts and 10 volts, the capacitor cannot change its voltage instantaneously; thus, the instant changes in the voltage V(1) appear across **R1**. We next obtain the **PROBE** traces of V(1) and V(2) obtained from a **Fourier analysis**.

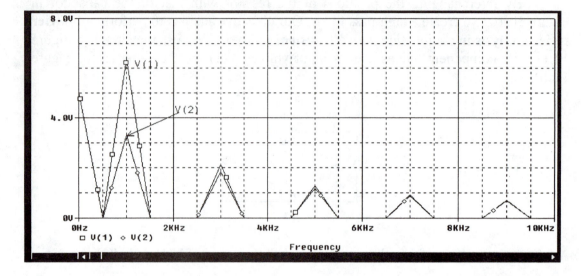

The trace of V(1) has a dc value of 5 volts, and the trace of V(2) has a dc value of zero volts. We can see this from their **transient traces** also. In particular, the voltage V(2) has zero net area under its curve for each of its cycles.   The fundamental harmonic for V(1) at over 6 volts is considerably larger that the fundamental harmonic for V(2), which measures about 3 volts. All the higher harmonics of the two voltages are practically identical. Such is the characteristic of a high-pass filter.

Some important points to remember: In the low-pass filter, the attenuation of the higher harmonics results in an increase of the rise time of V(2) compared to V(1). In the high-pass filter, the attenuation of the lower harmonics results in a decline of the dc value of V(2) compared to V(1).   We turn next to the **Output File** data.

```
FOURIER COMPONENTS OF TRANSIENT RESPONSE V(1)


DC COMPONENT =    4.950495E+00

HARMONIC   FREQUENCY      FOURIER     NORMALIZED     PHASE        NORMALIZED
   NO        (HZ)       COMPONENT    COMPONENT      (DEG)       PHASE (DEG)

    1      1.000E+03     6.366E+00    1.000E+00    -8.911E-01    0.000E+00
    2      2.000E+03     9.906E-02    1.556E-02    -9.178E+01   -9.000E+01
    3      3.000E+03     2.123E+00    3.334E-01    -2.673E+00   -2.551E-10
    4      4.000E+03     9.920E-02    1.558E-02    -9.356E+01   -9.000E+01
    5      5.000E+03     1.275E+00    2.002E-01    -4.455E+00   -1.266E-09
    6      6.000E+03     9.944E-02    1.562E-02    -9.535E+01   -9.000E+01
    7      7.000E+03     9.113E-01    1.431E-01    -6.238E+00   -3.504E-09
    8      8.000E+03     9.978E-02    1.567E-02    -9.713E+01   -9.000E+01
    9      9.000E+03     7.097E-01    1.115E-01    -8.020E+00   -7.393E-09
   10      1.000E+04     1.002E-01    1.574E-02    -9.891E+01   -9.000E+01


    TOTAL HARMONIC DISTORTION =    4.305759E+01 PERCENT
```

```
FOURIER COMPONENTS OF TRANSIENT RESPONSE V(2)


DC COMPONENT =  -5.804366E-02

HARMONIC   FREQUENCY    FOURIER    NORMALIZED    PHASE      NORMALIZED
  NO         (HZ)      COMPONENT   COMPONENT     (DEG)      PHASE (DEG)

   1       1.000E+03   3.340E+00   1.000E+00    5.703E+01   0.000E+00
   2       2.000E+03   1.057E-01   3.164E-02   -9.670E+01  -2.108E+02
   3       3.000E+03   1.852E+00   5.545E-01    2.529E+01  -1.458E+02
   4       4.000E+03   1.008E-01   3.018E-02   -9.742E+01  -3.256E+02
   5       5.000E+03   1.200E+00   3.592E-01    1.315E+01  -2.720E+02
   6       6.000E+03   9.953E-02   2.980E-02   -9.830E+01  -4.405E+02
   7       7.000E+03   8.775E-01   2.627E-01    6.472E+00  -3.928E+02
   8       8.000E+03   9.919E-02   2.969E-02   -9.954E+01  -5.558E+02
   9       9.000E+03   6.898E-01   2.065E-01    1.875E+00  -5.114E+02
  10       1.000E+04   9.927E-02   2.972E-02   -1.010E+02  -6.713E+02


   TOTAL HARMONIC DISTORTION =   7.434436E+01 PERCENT
```

By comparing the data for the voltage V(1) and V(2), we see that the former has a dc value of 5 volts, while V(2) has a dc value of zero volts. In corroboration with the **PROBE** data, the lower harmonics of V(2) are attenuated by comparison with those of V(1). The higher harmonics of both voltages are more nearly equal to each other.

Let us examine the action of this circuit on the harmonics of V(2). We shall in part use the previous tabulation of the frequency, the reactance of the capacitor, but a changed voltage-divider ratio as shown next. Again, we neglect the even harmonics due to their small amplitudes.

Frequency (kHz)	X_c (Ω)	V(2)/V(1) = R/(R-jX_c)
1	1590	$.53\angle 57.3°$
3	531	$.88\angle 27.9°$
5	318	$.95\angle 17.6°$
7	227	$.98\angle 12.8°$
9	177	$.98\angle 10.0°$

The ratio V(2)/V(1) is approaching the ratio **R/R = 1** as the reactance of the capacitor declines with the higher frequencies. Let us determine the amplitude of V(2) at the frequency of 1 kHz.

$$V(2)/V(1)_{(at\ 1\ kHz)} = .53\angle 57.3°$$

From which

$$V(2) = (.53\angle 57.3°)(6.37\angle -.89°) = 3.4\angle -56.4° \ V$$

The reader is encouraged to verify the voltage of V(2) for the other harmonics.

The Fall Time and the Critical Frequency

The **fall time** t_f of a pulse is defined as the time it takes for a pulse to go from 90% of its initial value to 10% of its final value. We recall our **PROBE** plot and measure the fall time using our cursors. The fall time measures 228.604 microseconds. This is identical to the **rise time** for the low-pass filter. This is to be expected since the same resistor and capacitor values were used in both the low-pass and high-pass filters. We recall that the critical frequency was a function of these two elements. Thus, both filters have the same identical **critical frequency** of 1.59 kHz.

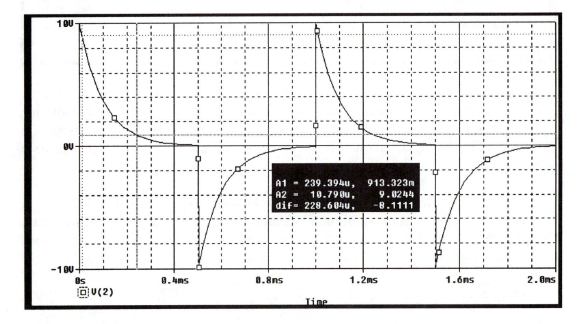

FOURIER SERIES ANALYSIS OF A TUNED AMPLIFIER

So far we have passed pulses through filters that have diminished the high frequencies or diminished the low frequencies. We now turn to a circuit that diminishes the harmonics at either extreme of the spectrum but emphasizes a band of them in the middle. Such a device is the tuned amplifier shown in Figure 10.12. We shall analyze its operations next. The gain of the dependent current source **G1** is set at .004.

Figure 10.12

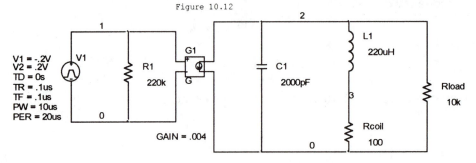

The **resonance frequency** of this tuned circuit is 250 kHz. The period of V(1) was set to 20 microseconds. Hence, the fundamental frequency is equal to 50 kHz. This makes the fifth harmonic of V(1) equal to the **resonance frequency** of the tuned amplifier. A **transient analysis** of 40 microseconds together with a **Fourier analysis** produced the frequency spectra of V(1) and V(2) and the **Output File** data.

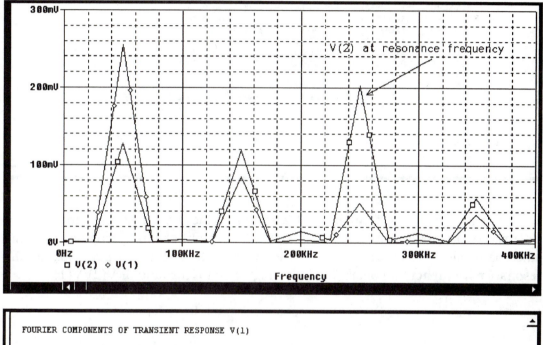

```
FOURIER COMPONENTS OF TRANSIENT RESPONSE V(1)

  DC COMPONENT =   2.000004E-03

  HARMONIC   FREQUENCY    FOURIER     NORMALIZED     PHASE        NORMALIZED
    NO         (HZ)      COMPONENT    COMPONENT      (DEG)       PHASE (DEG)

     1       5.000E+04   2.546E-01    1.000E+00    -1.800E+00     0.000E+00
     2       1.000E+05   4.000E-03    1.571E-02     8.640E+01     9.000E+01
     3       1.500E+05   8.482E-02    3.331E-01    -5.400E+00     9.229E-08
     4       2.000E+05   4.000E-03    1.571E-02     8.280E+01     9.000E+01
     5       2.500E+05   5.082E-02    1.996E-01    -9.000E+00     4.613E-07
     6       3.000E+05   4.000E-03    1.571E-02     7.920E+01     9.000E+01
     7       3.500E+05   3.623E-02    1.423E-01    -1.260E+01     1.291E-06
     8       4.000E+05   4.000E-03    1.571E-02     7.560E+01     9.000E+01
     9       4.500E+05   2.811E-02    1.104E-01    -1.620E+01     2.764E-06
    10       5.000E+05   4.000E-03    1.571E-02     7.200E+01     9.000E+01


   TOTAL HARMONIC DISTORTION =    4.295017E+01 PERCENT
```

```
FOURIER COMPONENTS OF TRANSIENT RESPONSE V(2)


DC COMPONENT =  -8.203522E-04

HARMONIC    FREQUENCY    FOURIER     NORMALIZED    PHASE      NORMALIZED
  NO          (HZ)      COMPONENT    COMPONENT     (DEG)      PHASE (DEG)

    1       5.000E+04   1.278E-01    1.000E+00   -1.512E+02   0.000E+00
    2       1.000E+05   3.300E-03    2.582E-02   -4.941E+01   2.531E+02
    3       1.500E+05   1.197E-01    9.370E-01   -1.397E+02   3.140E+02
    4       2.000E+05   1.145E-02    8.963E-02   -6.911E+01   5.358E+02
    5       2.500E+05   2.046E-01    1.601E+00    1.421E+02   8.983E+02
    6       3.000E+05   9.948E-03    7.785E-02   -1.679E+02   7.395E+02
    7       3.550E+05   5.819E-02    4.553E-01    8.930E+01   1.148E+03
    8       4.000E+05   4.893E-03    3.829E-02    1.725E+02   1.382E+03
    9       4.500E+05   2.712E-02    2.122E-01    7.884E+01   1.440E+03
   10       5.000E+05   3.307E-03    2.588E-02    1.653E+02   1.678E+03


   TOTAL HARMONIC DISTORTION =   1.926008E+02 PERCENT
```

At the **resonance frequency** of 250 kHz of the tuned amplifier, the trace of V(2) shows that its fifth harmonic has undergone amplification by contrast with the other harmonics. Specifically, if we use the data from the **Output File**, at the **resonance frequency**, we obtain the following value for the ratio of V(2)/V(1):

$$V(2)/V(1) \text{ (at resonance)} = .204/.05 = 4.1$$

Thus, the fifth harmonic of V(2) has been amplified by a factor of 4.1. The tuned amplifier is at the heart of the modern radio. A listener "tunes" to a particular station, usually by means of changing the capacitance of a tank circuit. If the radio is a high-quality device, only the tuned-to station will be heard. The frequencies of any adjacent station on the broadcast band will be sufficiently attenuated so as not to interfere with the signal of the tuned-to station.

PROBLEMS

10.1   Use the following circuit.

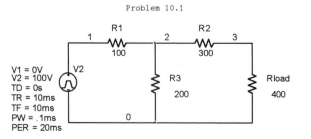

(a). From a **transient analysis** of 40 milliseconds duration, find the traces of V(1) and V(3); the latter is the voltage across **Rload.**

(b). Compare the appearances of the two traces.

(c). From your answer, can you predict the difference, if any, of the **Fourier** spectra of the two voltages?

(d). Perform a **Fourier analysis** and obtain the frequency spectra of the two voltages V(1) and V(2).

(e). Was your prediction from question (c) confirmed?

(f). Obtain the **RMS** power to **Rload** from the **Transient** data.

(g). Obtain the **RMS** power to **Rload** from the **Fourier** data.

(h). Compare the two **RMS** powers.

10.2 For this circuit, find the **Fourier spectra** of both the source voltage V(1) and that of the voltages V(1,2) and V(2). From a **Transient analysis** obtain the total **RMS** power into the circuit and the **RMS** power to each of the resistors. From the **Fourier data**, obtain the **RMS** power delivered by the voltage source V(1) and the **RMS** power received by each resistor. Compare the **RMS** powers from the two analyses. Are they the same?

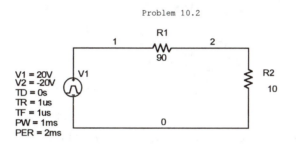

Problem 10.2

10.3 Use the following circuit.

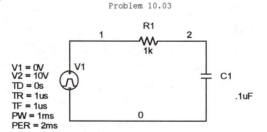

Problem 10.03

(a). Perform a **transient analysis** of 4 milliseconds duration and get the traces of V(1) and V(2).

(b). What is the maximum voltage of V(2)?

(c). Determine the rise time of V(2).

(d). Is this an integrator or differentiator circuit?

(e). Determine the **critical frequency** of this circuit.
(f). Perform a **Fourier analysis** to obtain the **Fourier spectra** of V(1) and V(2).
(g). What are the dc values of V(1) and V(2) respectively?
(h). Which are the strongest harmonics in V(1) and V(2)?
(i). Looking at the spectra, is this a low pass or high pass filter, relate this to your answer in question (d)?
(j). Which harmonics are present in V(1) and V(2), odd, even, or both?

10.4   Interchange the positions of the capacitor and the resistor in the previous problem. Answer all the questions as in that problem. In addition, state what remained the same and what changed.

10.5   For this circuit, perform a **Fourier analysis** for 10 harmonics of V(1) and answer the following:

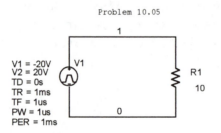

Problem 10.05

V1 = -20V
V2 = 20V
TD = 0s
TR = 1ms
TF = 1us
PW = 1us
PER = 1ms

V1

1

R1
10

0

(a). By inspection, what is the dc value of V(1)?
(b). Compare this with the dc value obtained from the **Output File** and the **PROBE** trace of V(1).
(c). What is the **RMS** value of each of the harmonics present?
(d). Find the **RMS** value of V(1).
(e). What is the power delivered to **R1** by each of the harmonics?
(f). Which one delivers the most, and which one delivers the least power?
(g). What is the percent contribution of each harmonic to the power to **R1**?

10.6   The voltage pulse **V1** of the previous problem has the **Fourier expansion** given in various textbooks as:

$$V(1) = -2V_m\left[\sin \omega t + \tfrac{1}{2} \sin 2\omega t + 1/3 \sin 3\omega t + \ldots + 1/n\sin n\omega t\right] \text{ volts}$$

Verify that the data from the **Output File** in the previous problem is in conformity with this expansion.

10.7   For the **RL** circuit shown, find the following:

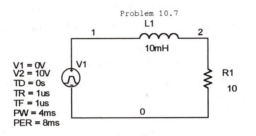

Problem 10.7

V1 = 0V
V2 = 10V
TD = 0s
TR = 1us
TF = 1us
PW = 4ms
PER = 8ms

L1
10mH

R1
10

Perform a **transient analysis** for 16 milliseconds.
(a). Obtain the traces of V(1), V(1,2) and V(2).
(b). What are the extreme values of the latter two voltages?
(c). What is the **rise time** of the V(2) pulse?
(d). From it, determine the **critical frequency** of this circuit.
(e). Compare the dc values of V(1), V(1,2) and V(2).

Perform a **Fourier analysis** using 10 harmonics.
(a). Obtain the dc values of V(1), V(1,2) and V(2).
(b). How do they compare to their values determined from the **transient analysis**?
(c). What are the amplitudes of the first harmonics of the three voltages?
(d). Which higher harmonics of what voltage decay the fastest?
(e). Obtain the **Output File** data for the three voltages.
(f). From it, calculate the **RMS** value of each voltage.
(g). What is the total **RMS** power delivered to **R1**?

10.8   The circuit has two voltage sources V1 and V3.

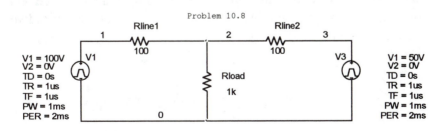

Problem 10.8

V1 = 100V
V2 = 0V
TD = 0s
TR = 1us
TF = 1us
PW = 1ms
PER = 2ms

Rline1
100

Rload
1k

Rline2
100

V1 = 50V
V2 = 0V
TD = 0s
TR = 1us
TF = 1us
PW = 1ms
PER = 2ms

(a). Find the **Fourier spectra** of the two voltage sources.

(b). What are the dc values of those voltages?

(c). What is the spectrum of the load voltage V(2)?

(d). What is its dc voltage?

(e). Which harmonics are present in all the spectra?

(f). What is the scale difference between the spectra of V(1) and V(3)?

(g). Compare the **harmonic distortion** of the three voltages. What can you conclude from your finding?

(h). Obtain the **RMS** power delivered by the two sources to **Rload.**

(i). Obtain the **RMS** power delivered to **Rload.** Compare this results to that of part (h).

10.9   This circuit is the same as in Figure 9.11, with the difference that **V1** now is a square pulse with the given parameters applied. You recall that the **resonance frequency** of that circuit was 1.41 kHz. For this problem, obtain the **Fourier spectrum** of V(2,3) and determine which of its harmonics are amplified. Relate this information to the **frequency analysis** that was performed in Chapter 9.

Problem 10.9

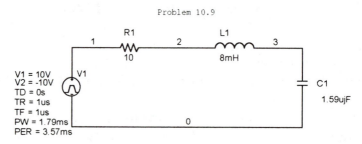

10.10   For the halfwave rectifier shown, find the dc value of the voltage V(2). Compare it to its theoretical value of V(1)/3.14. Which of the harmonics of V(2), aside from its fundamental, is most prominent? Obtain the trace of V(2) from a 1 millisecond **Transient analysis.** Is V(2) a half-wave rectified wave. What is its maximum amplitude? Why is its maximum amplitude different from that of V(1)?

Problem 10.10

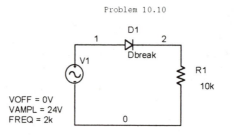

10.11   This is the circuit of Problem 10.10 but with its diode reversed.

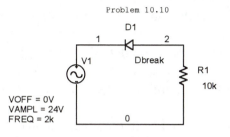

Answer all questions as in Problem 10.10. In addition, are the any differences in the time function of V(2) and its **Fourier spectrum** compared to V(2) of the previous problem?

10.12   The circuit in this problem is the same at that of Figure 9.14.

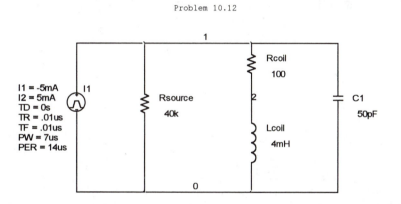

We now apply the current pulse shown. Note that period is 14 microseconds. This makes its fundamental frequency equal to 71.4 kHz and its fifth harmonic equal to 357.1 kHz. We recall that the **resonant frequency** of this circuit was equal to 354 kHz. Perform a **Fourier analysis** of this circuit and show that the fifth harmonic is the largest harmonic in the voltage V(1).

10.13   The circuit shown is a low-pass filter.

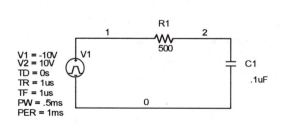

(a). Find the voltage pulses V(1) and V(2) as a function of time.

(b). Find the **rise time** of V(2).

(c). From this last data, determine the **critical frequency** of this filter.

(d). Obtain the frequency spectra of V(1) and V(2) for 10 harmonics.

(e). For each harmonic, calculate the ratio of V(2)/V(1).

(f). What happens to this ratio as the number of harmonics increases?

(g). How does this relate to the fact that our circuit is a low-pass filter?

10.14 The same circuit is used as in the previous problem. However, the period of **V1** has been changed to .4 milliseconds and the pulse width, to .2 milliseconds. For this condition, obtain traces of V(1) and V(2) as functions of time. Has the **critical frequency** of this filter changed? Obtain the new spectra of V(1) and V(2). Again, form the ratio of V(2)/V(1) for the 10 harmonics present. How do these ratios differ from the ones obtained in the previous problem and why?

10.15 The circuit in this problem is a full-wave rectifier.

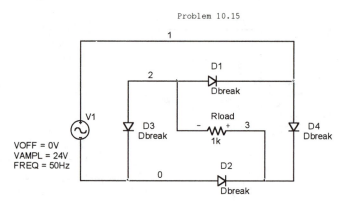

Problem 10.15

Perform a **Transient analysis** of 40 milliseconds duration and obtain the traces of the voltages V(1), V(2) and V(3,2). Do they conform to your expectations? Obtain the **RMS** power to **Rload**. Next, perform a **Fourier analysis** and obtain the traces of V(1), V(2) and V(3,2). What is the amplitude of the fundamental harmonic of V(1) and at what frequency does it occur? What is the **Fourier spectrum** of the voltage V(2) across the diode D3? What is its fundamental harmonic and what is its amplitude? What is its dc voltage? What harmonics are present in the spectrum of V(2)? What is the dc voltage of V(3,2)? Which is the largest harmonic of V(3,2) and what is its value? Which harmonics are present in the spectrum of V(3,2)? Compute the **RMS** value of the **Fourier expansion** of V(3,2) from its **Output File** data. Compare it to the **RMS** value obtained from the **transient analysis.**

10.16   The circuit shown is a two-section RC high-pass filter. In fact, it is the same circuit as in Figure 9.6. However, this time we shall perform a **Fourier analysis** and try to compare its data with that obtained from the previous **Frequency analysis.**

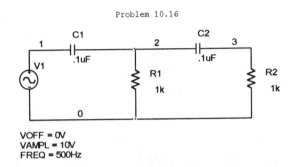

Problem 10.16

VOFF = 0V
VAMPL = 10V
FREQ = 500Hz

The fundamental frequency of V1 is set at 500 Hz. We need to run a **transient analysis** to obtain the data for the **Fourier analysis**. The setting of the duration of the **transient analysis** is left to the reader. From the **Fourier analysis**, on one plot, obtain the traces of V(1), V(2) and V(3). How many harmonics does each have and why? What is the frequency of the fundamental harmonic for each of them? What is the amplitude of the fundamental frequency for each of the voltages?

10.17   Perform a **Fourier analysis** of this circuit. From the **Output File** data file, write the **Fourier expansion** for V(1). Compare this expansion with one that has been published for the particular voltage waveform of V1.

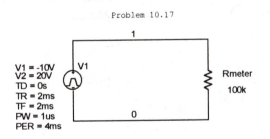

Problem 10.17

V1 = -10V
V2 = 20V
TD = 0s
TR = 2ms
TF = 2ms
PW = 1us
PER = 4ms

10.18   For the circuit shown, the **Fourier expansion** of V(1) is given in textbooks as:

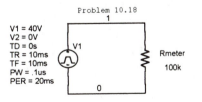

Problem 10.18

V1 = 40V
V2 = 0V
TD = 0s
TR = 10ms
TF = 10ms
PW = .1us
PER = 20ms

$$V(1) = \frac{V_m}{2} + \frac{4V_m}{(3.14)^2}\cos\omega t + \frac{4V_m}{(3)(3.14)^2}\cos 3\omega t + \frac{4V_m}{(5)(3.14)^2}\cos 5\omega t + \dots \quad \text{volts}$$

V_m = amplitude of **V1** = 40 volts

ω = fundamental radian frequency = 2*3.14*50 = 314 radians/second

Verify this expression for the above circuit by performing a **Fourier series** expansion of V(1).

10.19   The circuit in this problem is a modified version of the rectifier in Problem 10.10. A capacitor has been added in parallel with **R1** in an attempt to increase the dc voltage of the output voltage V(2). Run a **transient analysis** for two cycles of **V1** and note the effect on the trace of V(2). Can you explain the appearance of that trace? Just looking at it, do you think that the dc value of V(2) has increased over that of that voltage in Problem 10.19. Confirm your hunch and run a **Fourier analysis**. Obtain the **PROBE** traces of V(1) and V(2) and the harmonic contents of V(1) and V(2) as listed in the **Output File**.

        Compare the dc values of the voltages of V(2) of the present problem with that of Problem 10.10. Has there been the desired improvement? Compute the percent increase in the dc value. What other differences are there between the data for the two problems? Incidentally, the voltage V(2) in this problem is referred to as the filtered output voltage of the rectifier.

Problem 10.19

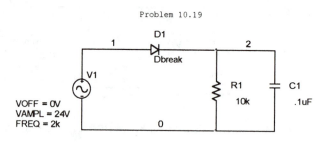

10.20   The circuit shown has the **VPWL** voltage source applied to it. Its voltage is
shown on the **PROBE** trace.   The reader is referred to Chapter 2 on the use of
the **VPWL** voltage source should such be necessary.

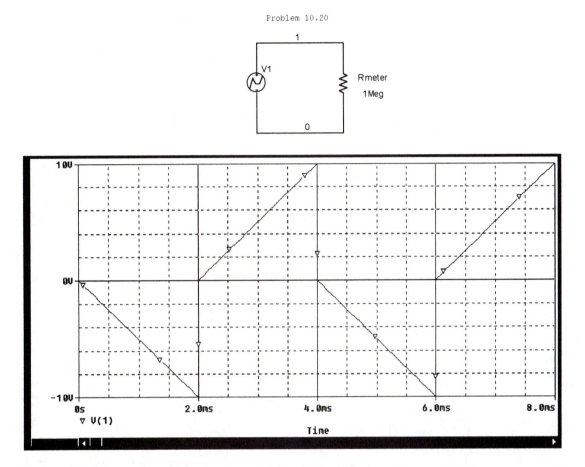

Write the **Fourier series** expansion of this circuit. The **transient analysis** parameters
are left for the reader to determine.

INDEX